Frontiers in Eating and Weight Regulation

Forum of Nutrition

Vol. 63

Series Editor

Ibrahim Elmadfa Vienna

Frontiers in Eating and Weight Regulation

Volume Editors

Wolfgang Langhans

ETH Zürich, Schwerzenbach

Nori Geary

ETH Zürich, Schwerzenbach

28 figures, 1 in color, and 4 tables, 2010

KARGER

Basel · Freiburg · Paris · London · New York · Bangalore · Bangkok · Shanghai · Singapore · Tokyo · Sydney

Wolfgang Langhans
Physiology and Behaviour Group
Institute of Food,
Nutrition and Health
ETH Zürich
Schwerzenbach, Switzerland

Nori Geary
Physiology and Behaviour Group
Institute of Food,
Nutrition and Health
ETH Zürich
Schwerzenbach, Switzerland

Library of Congress Cataloging-in-Publication Data

Frontiers in eating and weight regulation / volume editors, Wolfgang Langhans, Nori Geary.
p. ; cm. -- (Forum of nutrition, ISSN 1660-0347 ; v. 63)
Includes bibliographical references and indexes.
ISBN 978-3-8055-9300-7 (hard cover : alk. paper)
1. Appetite. 2. Body weight--Regulation. 3. Neuroendocrinology. 4. Gastrointestinal hormones. I. Langhans, Wolfgang. II. Geary, Nori. III. Series: Forum of nutrition, v. 63. 1660-0347 ;
[DNLM: 1. Eating--physiology. 2. Obesity--metabolism. 3. Adiposity--physiology. 4. Body Weight--physiology. W1 B1422 v.63 2010 / WD 210 F935 2010]
QP136.F76 2010
612.3'9--dc22

2009043618

Bibliographic Indices. This publication is listed in bibliographic services, including Current Contents® and PubMed/MEDLINE.

www.karger.com
Printed in Switzerland on acid-free and non-aging paper (ISO 9706) by Reinhardt Druck, Basel
ISSN 1660–0347
ISBN 978–3–8055–9300–7
e-ISBN 978–3–8055–9301–4

Contents

List of Contributors

William A. Banks
VAMC/St. Louis University School of Medicine
Internal Medicine, Geriatrics
915 Grand Boulevard
St. Louis, MO
USA

Denis G. Baskin
Department of Veterans Affairs
University of Washington
VA Puget Sound Health Care System
1660 South Columbian Way
Seattle, WA
USA

Rachel L. Batterham
Centre for Diabetes and Endocrinology
Department of Medicine
University College London
Rayne Building
5 University Street
London
UK

Christoph Beglinger
Division of Gastroenterology
University Hospital
Basel
Switzerland

James E. Blevins
Department of Veterans Affairs
University of Washington
VA Puget Sound Health Care System
1660 South Columbian Way
Seattle, WA
USA

Sebastien G. Bouret
The Saban Research Institute,
Neuroscience Program
Childrens Hospital Los Angeles
University of Southern California
USC Childhood Obesity Center
Keck School of Medicine
4650 Sunset Boulevard, MS#135
Los Angeles, Calif.
USA

Joanne E. Cecil
Bute Medical School
University of St Andrews
St Andrews
UK

Alain Dagher
Montreal Neurological Institute
McGill University
3801 University Street
Montreal, Quebec
Canada

Annette D. De Kloet
Program in Neuroscience
University of Cincinnati
2170 East Galbraith Road
Cincinnati, OH
USA

Nori Geary
Physiology and Behaviour Group
Institute of Food, Nutrition and Health
ETH Zürich
Schorenstrasse 16
Schwerzenbach
Switzerland

Kevin L. Grove
Oregon National Primate Research Center
Oregon Health & Science University
505 NW 185th Avenue
Beaverton, OR
USA

Marion Hetherington
Institute of Psychological Sciences
University of Leeds
Leeds
England

Jacquelien J. Hillebrand
Physiology and Behaviour Group
Institute of Food, Nutrition and Health
ETH Zürich
Schorenstrasse 16
Schwerzenbach
Switzerland

Morten L. Kringelbach
Department of Psychiatry
University of Oxford
The Queen's College
UK

Wolfgang Langhans
Physiology and Behaviour Group
Institute of Food, Nutrition and Health
ETH Zürich
Schorenstrasse 16
Schwerzenbach
Switzerland

Thomas A. Lutz
Institute of Veterinary Physiology
Vetsuisse Faculty University of Zürich
Winterthurerstrasse 260
Zürich
Switzerland

Timothy H. Moran
Department of Psychiatry and
Behavioral Sciences
Johns Hopkins University School of Medicine
Ross 618 720 Rutland Ave.
Baltimore, MD
USA

Heike Münzberg
Pennington Biomedical Research Center
Louisiana State University System
6400 Perkins Rd
Baton Rouge, LA
USA

Marianne T. Neary
Centre for Diabetes and Endocrinology
Department of Medicine
University College London
Rayne Building
5 University Street
London
UK

Alan Stein
Department of Psychiatry
University of Oxford
The Queen's College
UK

Gary J. Schwartz
Departments of Medicine & Neuroscience
Albert Einstein College of Medicine
1300 Morris Park Ave., Golding 501
Bronx, NY
USA

Eric Stice
Oregon Research Institute
1715 Franklin Boulevard
Eugene, OR
USA

E.L. Sullivan
Division of Neuroscience
Oregon National Primate Research Center
Oregon Health & Science University
505 NW 185th Avenue
Beaverton, OR
USA

Bettina Wölnerhanssen
Department of Visceral Surgery
University Hospital
Basel
Switzerland

Stephen C. Woods
Obesity Research Center
University of Cincinnati
2170 East Galbraith Road
Cincinnati, OH
USA

Preface

Scientific interest in the physiology of eating and body weight regulation has grown rapidly in recent years. There are both purely scientific and wider, cultural reasons for this development. The scientific reason relates to the advent of molecular genetics. The discovery of the adipose tissue hormone leptin by Jeffrey Friedman and his colleagues at Rockefeller University just 16 years ago revealed an important new neuroendocrine signaling pathway involved in the control of eating, energy expenditure and weight regulation and, more generally, made clear the power of molecular genetic techniques to help illuminate brain-behavior relationships. The influence, and the promise, of applying these tools to the study of eating and body weight regulation can hardly be overestimated. The cultural reason relates to the ongoing pandemic of obesity and of obesity-related health problems. The scale of the individual and societal costs of this pandemic have became clear only in the last 10–15 years. Unfortunately, equally clear is the current lack of effective strategies to control eating and body weight. The development of preventive and therapeutic options is a tremendous challenge to the science of eating and weight regulation in all its forms, from basic physiology to cognitive and social psychology.

Like previous advances in scientific technique and thought, the explosive growth in knowledge during the initial years of the molecular genetic revolution has been followed by a somewhat more intellectually critical phase, characterized by attempts to integrate new data and concepts with existing approaches. This is evident in the increasing numbers of studies in which cutting-edge molecular methodologies are combined with sophisticated traditional behavioral or physiological methods or with other new techniques, for example, functional imaging. In our view, the science of eating control and body weight regulation seems to be well into this synthetic period. As a result, the current scene is not dominated by a single type of methodology or single mode of thought. Rather, the wide boundary of the unknown is being pushed back in different ways and at different levels, often most successfully when different sorts of methods are combined.

Our book attempts to capture the spirit of this exciting era in the physiology of eating and weight regulation as well as its significance to the alleviation of the affliction of obesity. Together with the editors at Karger Publishers, we conceived a fresh approach to the usual volume of a collection of review articles. Our concept has two novelties: First, the main content of the book is a collection of brief, expert descriptions of recent developments in 15 examples of the important research frontiers in the physiology of eating, especially as it relates to weight regulation and adiposity. The intent of this format is to reflect several exciting recent developments in our area in an accessible form, so as to help inform and influence research in the coming years. To highlight the necessity of this continuing research, the book begins with a brief, expert introduction to the currently employed strategies for the treatment of obesity and their mostly disillusioning outcomes. The book's second unusual feature is that the frontier chapters are preceded by a general overview of the physiology of eating control and weight regulation. This overview chapter is meant both to provide requisite background information for the frontier chapters in an accessible way for readers for whom this is useful and to introduce an overarching conceptual and critical framework for the frontier chapters. As well, this chapter touches on several further active research areas that are not represented in the frontier chapters, such as new work on the satiating effect of glucagon-like peptide-1, advances in unraveling the complex role of brain serotonin in the control of eating, and the effects of bariatric surgery on physiological controls of eating and weight regulation, to name just three.

We hope that this approach has resulted in a book that is useful to students and newcomers to the field, to basic researchers engaged in the area, and to researchers and clinicians interested in the bidirectional translational dialog between bench and bedside. We have the optimistic view that the steady progress now visible in both basic and clinical research will generate increasingly effective treatments for disordered eating and body weight regulation. We hope that this book will facilitate this process. Last, but not least, we want to thank the editors at Karger Publishers for their patience and flexibility. Without their continuing support and understanding this book would not exist.

Wolfgang Langhans, Nori Geary
Schwerzenbach/Zürich

Langhans W, Geary N (eds): Frontiers in Eating and Weight Regulation.
Forum Nutr. Basel, Karger, 2010, vol 63, pp 1–8

Introduction – Obesity and Food Intake: Basic and Clinical Approaches

Annette D. De Kloet[a] · Stephen C. Woods[a,b]

[a]Program in Neuroscience and [b]Department of Psychiatry, University of Cincinnati, Cincinnati, Ohio, USA

Background

This introduction considers the current status of research on obesity and therapeutic strategies for it, including their relationships to the physiology of eating. Given the immense research effort currently targeting overweight and obesity, this summary is necessarily only a snapshot of a large and rapidly evolving area. It is nonetheless of immense importance since there is no sign that the obesity epidemic is abating, and because obesity per se carries so great a risk for numerous co-morbidities, such as type-2 diabetes mellitus (T2DM), several cardiovascular disorders and certain cancers. The topic is at the heart of the theme of this volume, given that obesity cannot exist unless energy intake (i.e. eating) chronically surpasses energy expenditure and since tackling aspects of eating represents, at least at present, the more approachable limb of the energy equation. As noted below, even the most successful therapeutic method now available, gastric bypass surgery, ultimately owes its efficacy to reduced energy intake.

Generally speaking, obesity refers to a state of excessive body fat and implies an unhealthy or undesirable body condition. Depending on one's perspective, obesity can be considered a symptom that carries an increased risk for numerous serious medical conditions or co-morbidities; a disease that warrants confrontation by governments, national health agencies, private benevolent groups, and third-party (health insurance) providers; or merely a warning that one should consider changing his or her lifestyle by consuming fewer calories each day [1]. Especially now that obesity has become a major focus of many health-care organizations, much new information has been forthcoming in the past few years and is beginning to influence the practice of medicine. It is important to realize that obesity is not a novel human condition; rather, evidence points to its existence in prehistoric times. What is novel is the persistent creep upward in the incidence of overweight and obesity in most human populations, a trend that is now widely considered an epidemic.

We now know much more about body fat than we did even a decade ago. Fat deposited in fat cells, or adipocytes, located in the abdominal region (i.e. the excess fat that increases waist circumference, whether subcutaneous or intra-abdominal) carries a greater risk for metabolic and cardiovascular disorders than fat located subcutaneously in the limbs or buttocks. As a general rule, females have a greater proportion of fat distributed subcutaneously whereas males have proportionally more abdominal fat. As fat mass increases, so does the complexity of the fat depot or individual fat organ; further, as obesity develops, both the size and ultimately the number of individual adipocytes increases. The increased fat mass is also associated with increased number and activity of macrophages and other immune cells that are attracted into the organ. These along with the adipocytes themselves secrete increasing amounts of hormones and other factors that predispose to metabolic and cardiovascular dysfunction, and they secrete less of some factors such as adiponectin that help prevent symptoms of diabetes. Several of these secretions are inflammatory factors, and obesity is now recognized as a chronic inflammatory disorder. Finally, as energy intake continues to outpace energy expenditure and body fat continues to expand, fat is deposited ectopically, i.e. outside the adipose tissue depots. Ectopic fat can occur in most tissues as obesity worsens, including the liver, heart, pancreas and skeletal muscle, and in each instance it compromises the normal functioning of those organs.

The increasing number of individuals with obesity, coupled with the growing understanding of the health risks obesity carries, has increased the urgency of developing safe and efficacious treatment options. The current therapeutic approaches for the treatment of obesity can be partitioned into lifestyle modifications, pharmacotherapy and bariatric surgery. The next sections briefly review each modality.

Lifestyle Modifications

Lifestyle modification is the first-order treatment for obesity recommended by the World Health Organization and the National Institutes of Health of the USA (NIH) [2, 3]. Their guidelines state that an individual should attempt lifestyle interventions for at least 6 months before other approaches are considered, and then to supplement the effort with additional approaches (e.g. pharmacotherapy) only with a physician's consent. Lifestyle interventions generally rely on increasing physical activity and/or decreasing caloric intake, with the goals of reducing body weight as well as decreasing the risk of the co-morbidities associated with obesity [4–6]. While this formula can be successful with frequent and intense educational and counseling programs, it is difficult for many obese individuals to maintain it for prolonged intervals without substantial support [5, 7]. Such relapse makes sense from a physiological perspective. That is, while significant weight can be lost in the short term (weeks or perhaps months), this recruits negative-feedback controllers, such as the adiposity signals discussed below, that work to thwart those efforts, with a common outcome being that most lost weight is regained within a year or two [7]. It must be asked, therefore, why gaining weight and becoming obese seems so much easier than being able to lose it. While there are no obvious answers to this apparent paradox, it does seem to be the case that the weight-regulatory system has an inherent bias favoring weight gain whenever the environment permits it [8]. Many people believe that the current epidemic of obesity is a natural consequence of an environment that favors taking in more energy (i.e. in the form of calorie-dense, palatable foods, or significant amounts of high-fructose corn syrup) while requiring less energy expenditure at many jobs, i.e. that it is an unhealthy lifestyle that leads to obesity in the first place.

Dieting is the most common approach adopted by people trying to lose weight. New popular diets appear regularly, most of which claim some unique advantage in helping individuals be successful [7]. Many entail increasing or decreasing the intake of one or another macronutrient (i.e. high or low proportion of fat, carbohydrate or protein). However, meta-analyses comparing the efficacy of such diets indicate that regardless of macronutrient composition, when matched for caloric content, the weight-reducing effects of popular diets are equipotent, i.e. macronutrient content is not important so long as caloric intake is less than caloric expenditure [7, 9].

Increased physical activity (i.e. more exercise) is considered an excellent alternative or complement to dieting, and it has the added benefit of improving other parameters, such as insulin sensitivity and muscle tone, independent of weight loss. Unfortunately, increasing exercise has proven to be even more difficult in the long-term than dieting for most obese or overweight individuals.

All in all, although lifestyle modifications are the initial and most common treatment options recommended for and used by overweight and obese individuals, their modest efficacy coupled with their poor long-term success has focused research efforts on other strategies, including pharmacotherapy and bariatric surgery.

Pharmacotherapy

Pharmacological targets for the treatment of excess weight include appetite (sibutramine), fat absorption (orlistat), weight-regulatory brain circuits (cannabinoid receptor-1 (CB1) antagonists), and metabolism (CB1 antagonists; drugs that stimulate uncoupling proteins). So-called 'off-label' applications of medications primarily intended for other illnesses, such as the antidepressant fluoxetine, also may facilitate weight loss. In addition, two types of medications targeting type-2 diabetes also have weight-lowering properties, GLP-1 agonists and amylin agonists. Nevertheless, only two compounds are currently approved for chronic weight loss in most countries: orlistat (Xenical, Roche Laboratories, Inc.) and sibutramine (Meridia, Abbot Labs, Inc.). Each results in an average weight loss of only 3–5 kg, and each has bothersome side effects, reducing long-term adherence. Given this situation, one readily comprehends the massive efforts of pharmaceutical firms and universities to exploit our understanding of the physiology of eating, as detailed throughout this book, to develop better medications for the treatment of obesity.

Sibutramine acts within the brain, reducing the reuptake of secreted serotonin and nor-epinephrine, and to a lesser extent dopamine [10]; hence, sibutramine necessarily impacts numerous circuits not directly relevant to energy homeostasis. Sibutramine reduces eating and may also elicit a small increase of energy expenditure [11]. Numerous clinical studies have documented the ability of sibutramine to cause weight loss and slow the rate of weight regain after dieting, as reviewed in recent meta-analyses [12, 13]. Chronic sibutramine treatment leads to modest weight loss, reduced body fat and waist circumference, and improved glycemic and lipid profiles. The major side effect is increased systolic and diastolic blood pressure and heart rate, symptoms that can be problematic in some individuals [11]. Although the average weight loss due to sibutramine is modest, an important point is that even small reductions of total fat translate into proportionally larger reductions of visceral or abdominal fat, the fat that poses the greatest risk for diabetes and cardiovascular problems [14].

Orlistat inhibits gastric and pancreatic lipase [15], resulting in about one third of ingested fat not being absorbed and consequently excreted in the feces [16]. A recent meta-analysis confirmed that orlistat reduces body weight, body fat, waist circumference and plasma glucose; results in slightly reduced systolic and diastolic blood pressure, and decreases plasma low-density lipoprotein (LDL) triglyceride [13]. The major side effect is oily fecal discharge, which greatly reduces long-term compliance. Direct clinical comparisons of sibutramine and orlistat suggest that sibutramine has a small, but significantly greater effect on weight loss and glycemic parameters.

Glucagon-like peptide-1 (GLP-1) is an intestinal incretin hormone secreted during meals, and increasing evidence indicates it plays a role in satiation [17, 18]. Because GLP-1 acts to augment prandial insulin secretion, small-molecule GLP-1 receptor agonists are prescribed as an adjunct treatment for T2DM. Patients receiving these compounds often experience modest weight loss in addition to improved glucose tolerance [19–21]. However, it is not clear how the compounds act to reduce weight because compounds that prevent the breakdown of endogenous GLP-1 share the antidiabetic but not the weight-lowering properties of GLP-1 agonists [20], and because the mechanism may not involve reduced food intake [19, 22]. Two other gut intestinal hormones that appear to have potential as antiobesity therapies are ghrelin and peptide YY [19].

Amylin is a peptide hormone co-secreted with insulin from pancreatic B cells, whose role in eating is reviewed elsewhere in this book [23] and by other authors [18, 24]. Amylin analogs are used in the treatment of diabetes, and can result in modest weight loss [23, 25].

CB1 receptor antagonists are another class of compounds with apparent promise for reducing body weight and improving glucose and lipid profiles. In both animal models and human clinical trials, CB1 agonists cause a transient reduction of food intake and maintained weight loss with associated reduction of plasma lipids and improved glucose tolerance [26]. In spite of the metabolic improvements, CB1 antagonists have not been widely approved as weight-loss agents due to a tendency to exacerbate mood disorders in some obese patients [27, 28]. An important goal of future research will be to develop analogs of these compounds that lack the undesirable side effects.

Bariatric Surgery

At present, the most efficacious treatments for reducing excess body weight are one or another type of bariatric surgery. These were initially developed with the intent to manipulate the gastrointestinal tract so as to alter the intraluminal capacity for food by reducing the volume of the GI tract, to reduce nutrient absorption, or both [29]. This led to procedures that place various kinds of restrictions to limit the available volume of the stomach into which swallowed food can enter (i.e. gastric bands or gastric sleeves) and/or rearranging the intestinal passageway so to reduce the transit distance covered by ingested food (e.g. roux-en-Y gastric bypass (RYGB); ileal interposition). The number of humans undergoing such procedures, and the number of variations of each procedure, has increased dramatically over the last few years, and new data are forthcoming regularly, such that any conclusions are likely to be modified over the next few years. A few generalizations can nonetheless be made, and most apply both to gastric banding and to RYGB, with RYGB having a greater effect in reducing body weight. First of all, the degree of weight loss achieved by bariatric surgery is dramatically greater than can be achieved by any presently known lifestyle or pharmacological means. Second, the weight loss is long-lasting in that many subjects have been followed for more than 15 years with little weight regain [30]. In addition, individuals with successful surgeries have reduced all-cause mortality over at least 15 years, pointing to a major health benefit [30]. Third, the major cause of weight loss seems to be reduced appetite and avoidance of fatty (i.e. energy dense) foods, with little evidence for malabsorption of nutrients. Fourth, and what has perhaps been the most surprising from the medical standpoint, is the reduction in the severity of symptoms of diabetes, with many bariatric surgery patients essentially undergoing complete remission at the time they are discharged from the hospital postsurgery and prior to significant weight loss [31, 32]. The mechanisms responsible for the decreased appetite and remission of diabetes are unknown, but probably include some combination of enhanced nutrient stimulation of the distal intestine and consequent enhanced release of incretin hormones (e.g. GLP-1), reduced stimulation of the proximal intestine, reduced secretion of gastric hormones such as ghrelin, or others [33].

Eating

This volume is rich with information on the myriad physiological influences on eating [17]. As a generalization, most factors that influence eating can be considered either homeostatic or nonhomeostatic, with homeostatic factors relating to the regulation of one or more key physiological parameters such as body fat, blood glucose, or energy availability. Nonhomeostatic influences include hedonic and emotional factors, learning and experience, the social situation, stress, circadian rhythms, and so on.

My colleagues and I summarized the organization of homeostatic factors a decade ago [34, 35], and the basic model still holds, albeit it with numerous refinements, many described in this volume, having being added. Thus, as described in more detail in another chapter of this volume [17], a few rudiments of the current view of the physiology of eating are: (1) the initiation of meals is most often due to non-homeostatic factors such as habit or convenience; (2) meal termination is determined in part by negative-feedback satiation signals such as cholecystokinin that are elicited during the meal, usually stimulate the hindbrain and act to increase the feeling of fullness and end the meal, and (3) hormones or other signals that are secreted in proportion to body fat (adiposity signals, such as insulin and leptin) are integrated at the level of the hypothalamus and alter the sensitivity of the brain to meal-generated satiation signals. Thus, if one is dieting and loses weight, adiposity signals are reduced and the brain becomes less sensitive to CCK and other satiation signals, and larger meals are consumed until body weight is restored. Conversely, excess weight gain is accompanied by increases in adiposity signals and the brain is more sensitive to satiation signals.

All aspects of this model are expertly covered in the various chapters of this volume. There are contributions on the generation and influence of satiation signals [19, 36], on adiposity signals and their entry into the brain [23, 37–39], on hypothalamic circuits [40], their sensitivity to nutrients [41, 42], and their interactions with the hindbrain [43, 44]. There are also contributions reviewing exciting new areas, including the articles by Cecil and Hetherington [45] and Neary and Batterham [46] on the role of genetic factors, Kringelbach and Stein [47] on the emerging field of functional brain imaging, and Stice and Dagher [48] on the integration of genetic and imaging approaches.

Conclusion

Although human and animal studies indicate that lifestyle modifications can be effective obesity therapies; indeed, as described in a recent study [49], they are sometimes more effective than pharmacological therapy, and the low level of adherence to these lifestyle therapies has focussed contemporary translational research for treating overweight and obesity onto pharmacological and surgical approaches. Considerable further research is both needed and ongoing in this regard. This volume makes a valuable contribution to providing the physiological foundation for that effort.

References

1 Smith GP: Critical introduction to obesity; in Blass EM (ed): Obesity: Causes, Mechanisms, Prevention, and Treatment. Sunderland, Sinauer, 2008.

2 National Institute of Health and National Heart Lung and Blood Institute: Clinical guidelines on the identification, evaluation, and treatment of overweight and obesity in adults – The Evidence Report. National Institutes of Health. Obes Res 1998;6(suppl 2):51S–210S.

3 World Health Organization: Obesity: Preventing and Managing the Global Epidemic. Geneva, WHO, 1998.

4 Fang J, Wylie-Rosett J, Alderman MH: Exercise and cardiovascular outcomes by hypertensive status: NHANES I epidemiological follow-up study, 1971–1992. Am J Hypertens 2005;18:751–758.

5 Wadden, TA, Butryn, ML, Wilson C: Lifestyle modification for the management of obesity. Gastroenterology, 2007;132:2226–2238.

6 Yamaoka K, Tango T: Efficacy of lifestyle education to prevent type 2 diabetes: a meta-analysis of randomized controlled trials. Diabetes Care 2005;28:2780–2786.

7 Dansinger ML, et al: Comparison of the Atkins, Ornish, Weight Watchers, and Zone diets for weight loss and heart disease risk reduction: a randomized trial. JAMA 2005;293:43–53.

8 Schwartz MW, et al: Is the energy homeostasis system inherently biased toward weight gain? Diabetes 2003;52:32–38.

9 Sacks FM, et al: Comparison of weight-loss diets with different compositions of fat, protein, and carbohydrates. N Engl J Med 2009;360:859–873.

10 Bray GA, Greenway FL: Pharmacological treatment of the overweight patient. Pharmacol Rev 2007;59:151–184.

11 Tziomalos K, Krassas GE, Tzotzas T: The use of sibutramine in the management of obesity and related disorders: an update. Vasc Health Risk Manag 2009;5:441–452.

12 Kim SH, et al: Effect of sibutramine on weight loss and blood pressure: a meta-analysis of controlled trials. Obesity 2003;11:1116.

13 Rucker D, et al: Long term pharmacotherapy for obesity and overweight: updated meta-analysis. BMJ 2007;335:1194–1199.

14 Despres JP, Lemieux I, Prud'homme D: Treatment of obesity: need to focus on high risk abdominally obese patients. BMJ 2001;322:716–720.

15 Carriere F, et al: Inhibition of gastrointestinal lipolysis by Orlistat during digestion of test meals in healthy volunteers. Am J Physiol Gastrointest Liver Physiol 2001;281:G16–G28.

16 Guerciolini R: Mode of action of orlistat. Int J Obes Relat Metab Disord 1997;21(suppl 3):S12–S23.

17 Langhans W, Geary N: Overview of the physiological control of eating; in Langhans W, Geary N (eds): Frontiers in Eating and Weight Regulation. Forum Nutr. Basel, Karger, 2010, vol 63, pp 9–53.

18 Woods SC, et al: Pancreatic signals controlling food intake; insulin, glucagon and amylin. Philos Trans R Soc Lond B Biol Sci 2006;361:1219–1235.

19 Wölnerhanssen B, Beglinger C: Therapeutic potential of gut peptides; in Langhans W, Geary N (eds): Frontiers in Eating and Weight Regulation. Forum Nutr. Basel, Karger, 2010, vol 63, pp 54–63.

20 Aulinger B, D'Alessio D: Glucagon-like peptide 1: continued advances, new targets and expanding promise as a model therapeutic. Curr Opin Endocrinol Diabetes Obes 2007;14:68–73.

21 DeFronzo RA, et al: Effects of exenatide (exendin-4) on glycemic control and weight over 30 weeks in metformin-treated patients with type 2 diabetes. Diabetes Care 2005;28:1092–1100.

22 Woods SC: The control of food intake: behavioral versus molecular perspectives. Cell Metab 2009;9:489–498.

23 Lutz TA: Roles of amylin in satiation, adiposity and brain development; in Langhans W, Geary N (eds): Frontiers in Eating and Weight Regulation. Forum Nutr. Basel, Karger, 2010, vol 63, pp 64–74.

24 Lutz TA: Control of food intake and energy expenditure by amylin-therapeutic implications. Int J Obes (Lond), 2009;33(suppl 1):S24–S27.

25 Aronne L, et al: Progressive reduction in body weight after treatment with the amylin analog pramlintide in obese subjects: a phase 2, randomized, placebo-controlled, dose-escalation study. J Clin Endocrinol Metab 2007;92:2977–2983.

26 de Kloet AD, Woods SC: Minireview: Endocannabinoids and their receptors as targets for obesity therapy. Endocrinology 2009;150:2531–2536.

27 Food and Drug Administration: FDA Briefing document. NDA 21–888. Zimulti (rimonabant) Tablets, 20 mg. Sanofi Aventis. Advisory Committee – June 13, 2007. Available at http://www.fda.gov/ohrms/dockets/AC/07/briefing/2007-4306b1-fda-backgrounder.pdf. Accessed August 9, 2007.

28 Isoldi KK, Aronne LJ: The challenge of treating obesity: the endocannabinoid system as a potential target. J Am Diet Assoc 2008;108:823–831.

29 Mun EC, Blackburn GL, Matthews JB: Current status of medical and surgical therapy for obesity. Gastroenterology, 2001;120:669–681.

30 Sjostrom L, et al: Effects of bariatric surgery on mortality in Swedish obese subjects. N Engl J Med, 2007;357:741–752.

31 Dixon JB, et al: Adjustable gastric banding and conventional therapy for type 2 diabetes: a randomized controlled trial. JAMA 2008;299:316–323.
32 Thaler JP, Cummings DE: Minireview: Hormonal and metabolic mechanisms of diabetes remission after gastrointestinal surgery. Endocrinology 2009; 150:2518–2525.
33 Cummings DE: Endocrine mechanisms mediating remission of diabetes after gastric bypass surgery. Int J Obes (Lond), 2009;33(suppl 1):S33–S40.
34 Woods SC, et al: Signals that regulate food intake and energy homeostasis. Science 1998;280:1378–1383.
35 Schwartz MW, et al: Central nervous system control of food intake. Nature 2000;404:661.
36 Langhans W: Peripheral metabolic signals; in Langhans W, Geary N (eds): Frontiers in Eating and Weight Regulation. Forum Nutr. Basel, Karger, 2010, vol 63, pp 75–83.
37 Banks WA: The blood brain barrier as a regulatory interface; in Langhans W, Geary N (eds): Frontiers in Eating and Weight Regulation. Forum Nutr. Basel, Karger, 2010, vol 63, pp 102–110.
38 Hillebrand JJG, Geary N: Do leptin and insulin signal adiposity?; in Langhans W, Geary N (eds): Frontiers in Eating and Weight Regulation. Forum Nutr. Basel, Karger, 2010, vol 63, pp 111–122.
39 Münzberg H: Leptin-signaling pathways and leptin resistance; in Langhans W, Geary N (eds): Frontiers in Eating and Weight Regulation. Forum Nutr Basel, Karger, 2010, vol 63, pp 123–132.
40 Bouret SG: Development of hypothalamic neural networks controlling appetite; in Langhans W, Geary N (eds): Frontiers in Eating and Weight Regulation. Forum Nutr. Basel, Karger, 2010, vol 63, pp 84–93.
41 Sullivan E, Grove K: Metabolic imprinting; in Langhans W, Geary N (eds): Frontiers in Eating and Weight Regulation. Forum Nutr. Basel, Karger, 2010, vol 63, pp 186–194.
42 Moran TH: Hypothalamic nutrient sensing; in Langhans W, Geary N (eds): Frontiers in Eating and Weight Regulation. Forum Nutr. Basel, Karger, 2010, vol 63, pp 94–101.
43 Blevins J, Baskin D: Hypothalamic-brainstem circuits controlling eating; in Langhans W, Geary N (eds): Frontiers in Eating and Weight Regulation. Forum Nutr. Basel, Karger, 2010, vol 63, pp 133–140.
44 Schwartz GJ: Brainstem integrative function in the central nervous system control of food intake; in Langhans W, Geary N (eds): Frontiers in Eating and Weight Regulation. Forum Nutr. Basel, Karger, 2010, vol 63, pp 141–152.
45 Hetherington M, Cecil JE: Gene-environment interactions in obesity; in Langhans W, Geary N (eds): Frontiers in Eating and Weight Regulation. Forum Nutr. Basel, Karger, 2010, vol 63, pp 195–203.
46 Neary M, Batterham R: Gaining new insights into food reward with functional neuroimaging; in Langhans W, Geary N (eds): Frontiers in Eating and Weight Regulation. Forum Nutr. Basel, Karger, 2010, vol 63, pp 152–163.
47 Kringelbach ML, Stein A: Cortical mechanisms of human eating; in Langhans W, Geary N (eds): Frontiers in Eating and Weight Regulation. Forum Nutr. Basel, Karger, 2010, vol 63, pp 164–175.
48 Stice E, Dagher A: Genetic variation in dopaminergic reward in humans; in Langhans W, Geary N (eds): Frontiers in Eating and Weight Regulation. Forum Nutr. Basel, Karger, 2010, vol 63, pp 176–185.
49 Diabetes Prevention Program Research: Reduction in the incidence of type 2 diabetes with lifestyle intervention or metformin. N Engl J Med 2002;346: 393–403.

Stephen C. Woods
Department of Psychiatry, University of Cincinnati
2170 East Galbraith Road
Cincinnati, OH 45237 (USA)
Tel. +1 513 558 6799, Fax +1 513 297 0966, E-Mail steve.woods@uc.edu

Langhans W, Geary N (eds): Frontiers in Eating and Weight Regulation.
Forum Nutr. Basel, Karger, 2010, vol 63, pp 9–53

Overview of the Physiological Control of Eating

Wolfgang Langhans · Nori Geary

Physiology and Behaviour Group, Institute of Food, Nutrition and Health, ETH Zürich, Schwerzenbach, Switzerland

Introduction

Aim

In this chapter we discuss the physiology of eating, with a particular focus on its relevance to the present obesity epidemic. The physiology of eating comprises the functional organization of eating behavior, the types of exteroceptive and interoceptive information that affect eating, the neural and endocrine sensory mechanisms relaying this information to the central nervous system (CNS), and the CNS neural networks that process and integrate this and other information to control eating (fig. 1). We emphasize the role of eating in the regulation of body weight. These topics have taken on new importance with the obesity epidemic. It is well recognized that overeating together with reduced exercise are the proximal causes of obesity. Therefore, better understanding of the physiology of eating and its role in body weight regulation, or dysregulation, should lead to new and hopefully more effective approaches for the therapeutic control of eating in obese persons or persons at special risk for obesity and obesity related diseases.

The Functional Organization of Eating

Eating in humans and other mammals is functionally organized into discrete meals. Meals are produced by four separable functional processes with at least partially independent underlying neural mechanisms. Although each process includes both behavioral and subjective phenomena, for simplicity we use single names for both aspects. The four processes are: (1) processes related to the initiation of meals (*hunger*

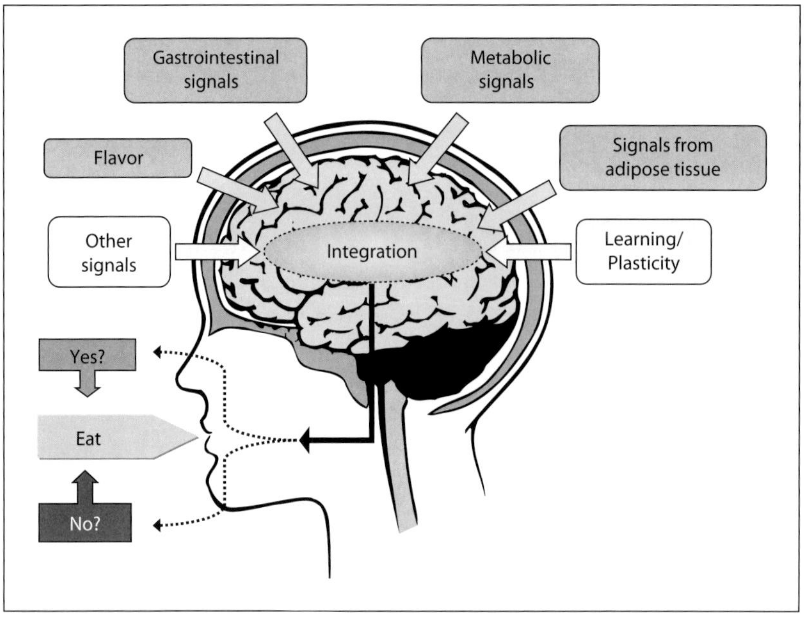

Fig. 1. Schematic of the control of eating. The decision to eat or not eat that is made before each bite or sip is the outcome of central nervous system *integration* of a variety of peripheral signals, including peripheral neural, hormonal and other humoral signals, information stored in the brain, such as learned effects of previous experience, food expectancies, etc., and with other signals, such as circadian or immune effects, situational context, energetic demands, etc. The schematic is superimposed on a shadow drawing of a midsagittal section of the head, including the skull, brain and spinal column. Reproduced with permission from Langhans et al. [280].

processes); (2) processes related to the evaluation of the food that stimulate or inhibit eating during the meal; this is one aspect of *food reward*; (3) processes related to inhibitory feedbacks from postingestive food stimuli that act to terminate eating at the end of the meal *(satiation)*, and (4) processes inhibiting eating during the intermeal interval *(postprandial satiation)*. As will become clear, each of these processes is affected both by phasic inputs, for example, inputs related to the secretion of hormones from the gut before, during and after meals, and by tonic inputs, for example, inputs related to the mass of the adipose tissue and, therefore, body weight.

An important implication of the fact that at least partially separate mechanisms control different meal processes is that summary measures of food intake, e.g. g/day, may conflate independent underlying processes. For example, in some situations, meal size and meal frequency change in opposite ways, so that the patterns of spontaneous eating can be different even though total amount eaten is not. A related point is that parts of the overall eating-control neural network with functionally antagonistic effects may operate simultaneously. Thus, eating-inhibitory controls might arise in one part of the network (e.g. homeostatic signals related to metabolic fuel utilization) at the same time as eating-stimulatory controls are activated in other parts of the network (e.g. signals related to orosensory food reward). The existence of such partially autonomous controls may be part of the reason why existing treatments based on pharmacological manipulation of single signaling molecules have not been effective in normalizing disordered eating.

Regarding the subjective phenomena associated with eating, our view, following William James [1], is that the most parsimonious explanation for the richness of eating-related emotional and cognitive experiences is that they evolved as causal agents contributing to the overall control of eating and its orchestration with other biological functions. How conscious processes actually affect neuronal function and behavior is, of course, beyond the scope of available methodologies, although imaging methods now produce at least hints that eating behavior and some of the subjective phenomena associated with eating arise in the same neural networks and are modulated by the state of energy balance. The chapters by Neary and Batterham [2], Kringelbach and Stein [3] and Stice and Dagher [4] touch upon this fascinating topic.

There are levels of organization of eating behavior both above and below the levels of meals. Subordinate to the level of meals is the microstructure of eating, including, for example, analyses of licking, biting, chewing or swallowing food during meals. This level of analysis seems to hold great potential for tracking eating behaviors via the lower motor neurons and central pattern generators that produce the movements of eating back into the higher, more integrative levels of the neural networks for eating. Superordinate levels include the control exerted by biological rhythms, such as circadian and reproductive rhythms, both of which potently affect human eating. The most important superordinate level, however, is the level mediating the regulation of body weight. That is, when adiposity or body weight is perturbed, the organism tends to eat in a way that corrects the error. The physiology of such weight-regulatory influences is a major theme of this book and is introduced in more detail in the next section.

Eating and Homeostasis

Because eating is our only source of metabolic fuel and of a number of essential nutrients, it is an integral part of homeostatic regulation. Myriad studies have demonstrated that both the regulation of metabolic energy supply and the regulation of micronutrient balance powerfully influence how much is eaten and what is eaten. As a consequence, homeostasis is a major conceptual scheme used to understand eating.

Body weight, at least in adulthood, is a relatively accurate surrogate for the state of energy balance over longer periods (i.e. periods during which changes in gut contents, hydration, etc., can be ignored). Over such periods, changes in body weight in adults usually reflect changes in adiposity, i.e. mainly the amount of energy substrate stored as triacylglycerols in the adipose tissue (there is also ectopic storage of

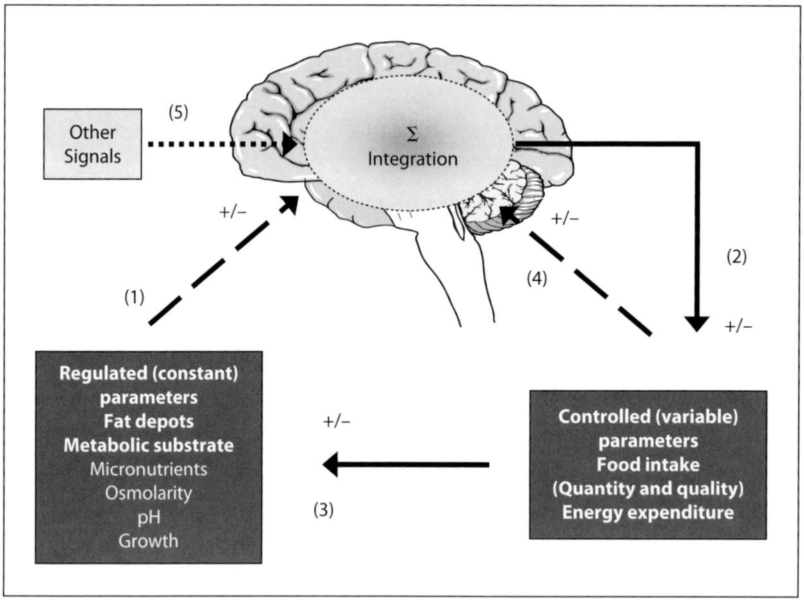

Fig. 2. Schematic of the components of homeostatic regulatory systems involved in the control of eating. Regulated parameters (lower left box) are held relatively constant, in part by changes in controlled or variable parameters (lower right box). The negative feedback control system thought to regulate body adiposity is shown in bold font; other regulated variables controlling eating are shown in normal font. In adiposity regulation: (1) Feedback signals reflecting deviations from the desired value (set point) in the regulated parameter, adiposity, are detected by the brain, (2) causing compensatory changes in eating or energy expenditure, which (3) affect adiposity. In addition, (4) eating produces other feedback signals to the brain that affect the control of meal onset, rate of eating, and meal termination. Finally, (5) other exogenous and endogenous signals outside these feedback loops also affect eating. Modified with permission from Langhans et al. [281].

triacylglycerols in liver, muscle and other tissues). Thus, longer-term state of energy balance is described by the energy balance equation:

Energy stored = Energy ingested – energy expended.

The relative stability of body weight over longer periods appears possible only if an active regulatory system senses energy stored and, depending on its level, appropriately adjusts energy ingested or energy expended. Figure 2 depicts how this system is believed to function. The brain registers and integrates (Σ) feedback signals which reflect deviations from the desired state (1), and adjusts eating and energy expenditure (the controlled variables) (2), so that the regulated variable, energy stored, is maintained in a relatively narrow envelope (3). The regulation of energy homeostasis and body weight is discussed further in this chapter as well as in the chapter by Hillebrand and Geary [5].

As also shown in figure 2, the feedback signals that are not related to energy homeostasis also affect the controlled variables. These include signals related to the sensory properties (especially, food palatability), volume and composition of the food (4). In addition, signals that are not related to the regulated or the controlled variables can also influence the system (5). Under certain circumstances, these latter two categories of signals can substantially disrupt regulation. There is little doubt that a main cause of the obesity epidemic in developed countries is the easy availability of increasingly palatable and energy-dense foods together with the decreased need (or opportunity) to exercise – for weight regulation, an 'obesifying' or 'toxic' environment.

Finally, several types of systems can produce regulation, and not all of them include the same components shown here. Whether energy homeostasis or other regulations affecting eating include reference values, or set points, as shown in figure 2, or whether constancy results from equilibria among feedback mechanisms without reference values, remains a matter of active debate.

Orosensory Signals in the Control of Eating

Flavor is a complex perception that arises from olfactory, gustatory, tactile and thermal food stimuli affecting receptors in the oro-nasopharynx. This sensory information can control eating independent of other pre- or postabsorptive consequences of eating, although association with such consequences normally determines much of the functional meaning of flavor stimuli.

The first type of process through which flavor affects eating is discrimination. This refers to flavor's informational content, i.e. identification of the type (e.g. 'it's sweet') and intensity ('it's as sweet as candy') of food stimuli, *independent* of the stimulus' rewarding qualities described below. Discriminative processes enable flavor stimuli to contribute to eating-related associative learning, which is important for both physiology (e.g. cephalic phase gastric and endocrine reflexes) and behavior (flavor-cued food selection, conditioned hunger and satiation).

The second type of process to which flavor stimuli contribute is reward. 'Food reward' is used to describe three potentially distinct ways in which flavor stimuli can influence eating: (1) Positive and negative flavor feedbacks that stimulate or inhibit ongoing eating. These can be either unconditioned or conditioned, are relatively automatic or reflexive, and are potent controllers of meal size. (2) Flavor hedonics, i.e. the pleasant or unpleasant subjective experiences of food stimuli ('I like sweet'), which are also thought to be sufficient to affect eating. (3) The reinforcing properties of flavor stimuli, meaning the sufficiency of flavor alone to produce long-term learned changes in behavior [3, 6].

The neural processes producing flavor hedonics are mainly cortical, based in part in specialized cortical regions which receive inputs from gustatory, olfactory and other senses (fig. 3). This makes food reward especially amenable to functional imaging techniques, as exemplified in three chapters [2–4]. Although we experience the effects of positive and negative feedbacks on eating and food palatability simultaneously, neural analyses indicate that these are often independent, separable processes.

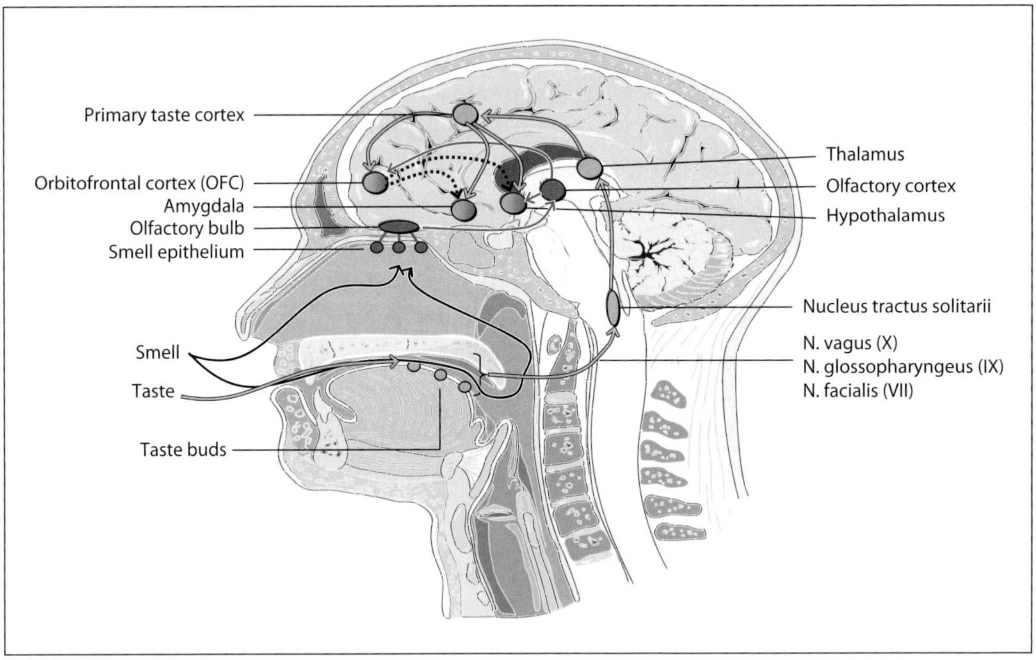

Fig. 3. Schematic showing the principal central projections of the gustatory and olfactory systems and their convergence in the hypothalamus, amygdala, and orbitofrotal cortex. See text for further details.

The direct effect of flavor on ingestion can be demonstrated in rats that sham feed with open gastric cannulas, which prevents significant accumulation of food in the stomach or entry of food into the duodenum (fig. 4) [7]. Similar tests can be done in humans by instructing subjects to take food into the mouth, to chew, etc., normally, but to spit it out rather than swallowing it [8]. In sham-feeding rats, the rate of ingestion varies directly with the concentration of preferred flavors, such as sugar or oil, and inversely with the concentration of nonpreferred flavors, such as bitter or salt.

Other terms and concepts are also used in the analysis of flavor's effects on eating. The terms 'palatability' and 'preference and aversion' are very common, and palatability has recently been further divided into 'wanting' and 'liking' processes [9]. 'Incentive reward' and 'craving' are also often discussed. All of these terms have been applied to both human and animal research. The extent to which they reflect different functional categories with different underlying physiological mechanisms remains an experimental question.

Although some preferences (sweet) and aversions (bitter, sour) for basic gustatory stimuli appear to be innate, preferences and aversions for the vast majority of flavors are learned. Gastrointestinal and postabsorptive consequences of the food can reinforce such learning [10]. This occurs in conditioned satiations, conditioned aversions (including the marked aversions for flavors associated with acute upper

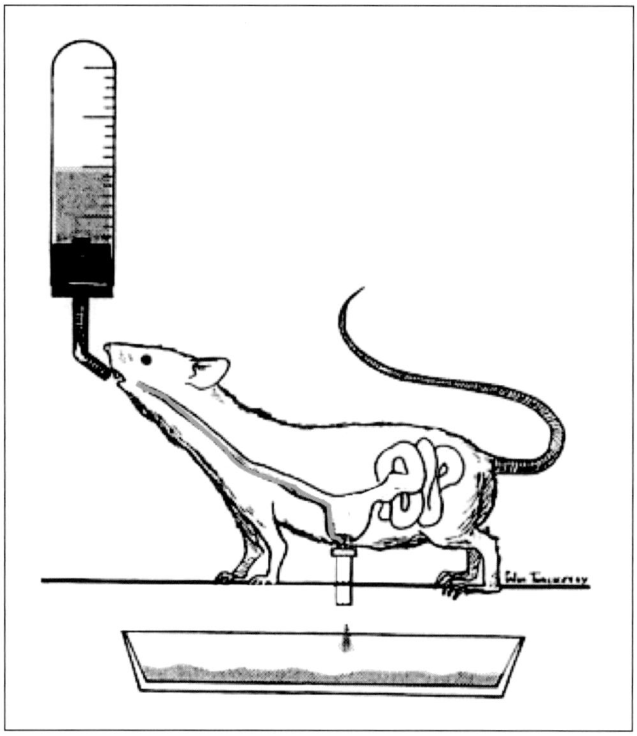

Fig. 4. Sham feeding in a rat equipped with a chronic gastric cannula, which is opened during sham feeding tests and closed for normal eating. During sham feeding, ingested food drains from the stomach. Modified with permission from Liebling et al. [282].

gastrointestinal illness), and 'specific hungers' (preferences for flavors associated with foods containing vitamins or minerals that can be learned during states of nutritional deficiency; this occurs for most micronutrients) [11]. In these situations, it is the discriminative, i.e. non-hedonic, aspects of the flavors that are important for learning, and increases or decreases in flavor hedonics are part of what is learned. The majority of human flavor preferences, however, are based not on physiological consequences of eating but on emotional, cognitive, and cultural associations attached to various foods, independent of their nutritional or physiological properties [10, 12, 13]. Indeed, mere exposure, i.e. familiarity, is sufficient to condition flavor preferences. This phenomenon likely explains much of the marked cultural variety in which foods are preferred, the social contexts or times of day when they are eaten, etc. [14] (and perhaps the preference for variety considered below). Because they dramatically affect patients' success in adhering to therapeutic dietary regimens, the origins and plasticity of human food preferences are important areas for behavioral and physiological research.

The increased availability of highly palatable foods in our society is considered a main cause of the increased prevalence of obesity. Consistent with this, differences in the palatability of both sweet and fat flavors have been shown between thinner and heavier humans [15–17]. Genetic variation in human flavor processing may also contribute to obesity. Obese individuals seem to be both less sensitive to the sensory intensity of sweet flavors and to enjoy both sweet and fat flavors whose sensory intensity is matched more than nonobese individuals do [15]. Furthermore, otitis media, a common childhood ear infection, can produce lifelong changes in flavor perception if the infection involves the trigeminal and glossopharyngeal nerves, which lie near the middle ear. Both children and adults with histories of severe otitis media have been reported to prefer sweets more than the general population and are at higher risk for overweight or obesity [15].

Variety is an important contributor to palatability. In both rats and humans, offering a variety of nutritionally identical foods with different, preferred flavors leads to larger meals than does offering only one of the alternatives, even the single most preferred one. The decrease in meal size when only one flavor is offered is referred to as sensory-specific satiety [18]. Flavor variety has also been shown to increase intake in the longer term in rats, leading to increased body weight [19].

Gastrointestinal and Pancreatic Signals in the Control of Eating

Introduction

The gastrointestinal (GI) system, pancreas and liver cooperate in the digestion and absorption of ingested food. A wide variety of physiological signals controlling eating also arise in these organs. In this section we describe what are classically considered *preabsorptive* GI signals. The next section, on metabolic controls, focuses on *postabsorptive* signals, which arise in the liver and outside the gut. This division, however, is only heuristic and organizational. For example, as described below, some pancreatic hormones are also released in the first minutes of eating via neuroendocrine reflexes and contribute to satiation, and we discuss these here as well. In addition, as considered in the next section, recent data suggest that metabolic controls of eating may also arise within the GI system, in the intestinal epithelia.

The GI system and the brain communicate via chemical and neural signals (fig. 5). The chemical signals include GI and pancreatic peptides whose release is affected by eating. Secretion of all but one of these, ghrelin, increases during and after meals. Ghrelin secretion, in contrast, increases during intermeal intervals. Neural signals include vagal and spinal visceral afferents originating in the gut.

Because of the important role of chemical messengers in the control of eating, it is useful to review some of the basic aspects of this sort of chemical signaling. Many GI chemical signals involved in the control of eating have a classical endocrine mode of action, i.e. specialized cells synthesize the signal molecule and in response to particular stimuli secrete it into the extracellular space, from which it diffuses into local capillaries, travels in the blood to a distant site, and binds to specific receptors that initiate its biological action. Some GI chemical signals, however, have a paracrine mode of

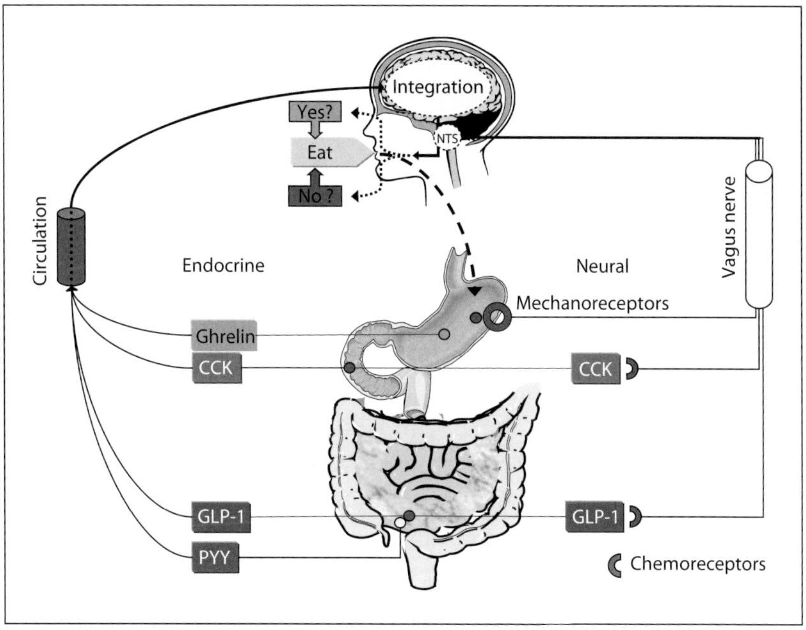

Fig. 5. Schematic of some important GI controls of eating. These act on the brain through neural (right) and endocrine (left) routes, as described in the text. Neural receptors: C = chemoreceptors; M, mechanorecpetors; Hormones: CCK = cholecystokinin; GLP-1 = glucagon-like peptide-1; PYY = peptide YY. Modified with permission from Langhans et al. [280].

action, which differs in that the signal molecule acts locally, reaching the target cells before entering the blood. Some signal molecules seem to have both modes of action. In addition, circulating levels of GI chemical signals are often many times higher in the hepatic portal vein than in the general circulation, which may be an important consideration when assessing the physiological actions of GI signals that act locally or in the liver. Another complexity arises in the case of endocrine signals that act in the brain to affect eating. Because of the selective barrier and active transport features of the blood-brain barrier (BBB), brain levels of hormones and metabolites are not simple mirrors of plasma levels. This issue is taken up in the chapter by Banks [20]. Finally, in the case of most gut hormones (ghrelin, cholecystokinin = CCK, glucagon-like peptide-1 = GLP-1, etc.) the same molecule is also synthesized by CNS neurons and acts as a neurotransmitter, often with a role in eating. This greatly complicates the interpretation of the phenotypes of mice with global null mutations (knockouts) of the molecule or its receptors.

Endocrine signals, because they appear in the systemic circulation, have been especially intensively investigated. This work has often utilized sets of explicit empirical criteria, modeled on classic endocrinological concepts, for the determination of which endogenous endocrine signals are normally involved in the control of eating, i.e. play *physiological* and not just *pharmacological* roles [21–23]. Evaluation of pharmacological signals is of course also important, as therapeutics can be based on either physiological or pharmacological actions of particular signals. The two major endocrine criteria for physiological function are called the physiological dose criterion and the antagonist criterion. The former is that administration of the hormone in amounts that mimic the endogenous (physiological) changes that occur at its site of action related to eating should be sufficient to produce the hypothesized effect on eating. The latter is that acute antagonism of the endogenous hormone at the time of its action on eating should reverse the effect. It is important that the antagonism be acute because physiological systems react to chronic manipulations, so the result of chronic antagonists is often to reveal active compensatory responses rather than essentially normal function except for one missing signal. This is another reason that complicates the physiological interpretation of the phenotypes of transgenic animals with global null mutations of specific genes. The sections below introduce some of the GI signals that at present appear to be particularly important in the physiological control of eating (for more detailed reviews, see [21, 24–29]). The therapeutic potential of several gut hormones is discussed in this volume by Wölnerhanssen and Beglinger [30].

Ghrelin

Ghrelin (fig. 5), a hormone discovered in 1999 [31], is the endogenous ligand for the growth hormone secretagogue receptor (GHS-R). Ghrelin is synthesized and secreted mainly by gastric X cells, but also by neurons in the CNS and other tissues. Gastric ghrelin has attracted great interest because it is the only gut peptide whose secretion is stimulated during fasting and inhibited by eating, and because it is the only gut peptide whose administration stimulates eating, which has been shown in rats and humans [21, 31–33].

The physiological status of ghrelin is not fully established. For example, it is unknown whether mimicking physiological ghrelin levels, especially the physiological pre-prandial rise in circulating ghrelin, is sufficient to trigger eating. GHS-R antagonists have been reported to decrease eating, but their selectivity remains uncertain [34]. An interesting alternative approach is the use of specific ghrelin spiegelmers [35, 36], which have recently been shown to reduce weight gain in mice offered a high-fat diet [37]. Another promising therapeutic approach related to ghrelin is based on pharmacological antagonism [38–40] of the recently discovered enzyme ghrelin O-acyltransferase [41], which catalyzes production of the biologically active acylated form of ghrelin.

The site of ghrelin's eating-stimulatory action is controversial. Some reports suggest that ghrelin acts peripherally to generate a vagal signal [42, 43]. More recent work, however, indicates that the eating-stimulatory effect of ghrelin does not require vagal afferent signaling [44]. Activation of GHS-R in the brain, especially those in the hypothalamic arcuate nucleus (Arc) and the brainstem, are sufficient to stimulate eating and are a likely mechanism for the endogenous eating effects of ghrelin [44, 45]. Neurons in these areas also synthesize and release ghrelin, and the relative contributions of hormonal and neuronal ghrelin on eating have not yet been distinguished.

Gastric Mechanoreception

The stomach is richly innervated with mechanoreceptors (fig. 5) that respond during and after meals and that signal the brain via both vagal and splanchnic visceral afferents. The effects of gastric mechanoreceptor signaling on eating have been studied in relative isolation in rats equipped with gastric cannulas, from which fluids can be infused or drained from the stomach, and pyloric cuffs, which can be inflated to prevent food from entering the intestines [46–48]. These experiments indicate that: (1) when gastric cannulas are used to prevent ingested liquid food from accumulating in the stomach, meal size is dramatically increased; (2) when ingested food is prevented from entering the intestines by inflating pyloric cuffs, meal size is about normal; (3) when fluid loads are infused into the stomach of rats with closed pyloric cuffs, eating is inhibited in proportion to the volume infused, and (4) the effect of gastric fill on eating is identical whether nutrient or non-nutrient loads are used. This indicates that gastric volume is an adequate stimulus for mechanoreceptors that can contribute to the control of eating. These signals, however, do not appear sufficient for the normal control of meal size in rats because intragastric infusions inhibit eating in rats with closed pyloric cuffs only when the total gastric fill (ingesta plus infusion) is markedly larger than the control meal size.

The pyloric cuff model does not fully assess the contribution of gastric mechanoreception to the control of eating. In both rats and rhesus monkeys, the intrameal rate of gastric emptying of liquid diet is about five times the postmeal rate [47]. As described above, the prevention of normal intrameal gastric emptying in the cuff-closed condition produces abnormal increases in gastric volume at meal end. It also prevents any interaction between gastric and postgastric signals. Many data indicate that such interactions are normally important; some examples are described in the chapter by Schwartz [49]. Thus, although gastric signals may not be sufficient for the control of meal size, they may indeed contribute importantly.

The role of gastric signals has also been studied in humans. Inflation of a gastric balloon before meals increases feelings of fullness and reduces meal size in normal-weight and obese subjects [50, 51]. The crucial signal may be related to fill of the antrum rather than fill of the fundus because sonographically measured antral cross-sectional areas after meals, but not fundal areas, correlated with fullness at meal end [52] and with the size of the next meal [53]. When the antral area was increased with a balloon before, but not during, the test meal, however, similar volumes had no effect on eating [54]. This may reflect a crucial role for interactions between gastric volume and postgastric food stimuli to elicit satiation, although Oesch et al. [54] were not able to detect such an interaction with satiating intraduodenal fat infusions (this method is described in the next section). Finally, a recent imaging study suggests that perceptions of fullness arising from increased gastric volume involve the amygdala and the insular cortex [55].

Intestinal Cholecystokinin

Cholecystokinin (CCK) (fig. 5) secreted mainly from duodenal I cells during and after meals has long been considered an essential physiological control of gastric emptying, gall bladder emptying, and exocrine pancreatic secretion. The classic report of Gibbs et al. [22] that intraperitoneal injections of CCK selectively inhibit eating established satiation as another potential physiological function of CCK, and CCK has remained the paradigmatic gut peptide eating-control signal. CCK was the first gut peptide whose satiating action fulfilled the criteria described above for a physiological control of eating in humans [21, 56–58]. There are two reports that increases in CCK mimicking prandial levels are sufficient to inhibit eating in humans [59, 60], supporting the physiological dose criterion described above. There are also, however, several reports that near physiological doses do not affect eating (moderate pharmacological doses, in contrast, decrease eating in humans without subjective or physical side effects). One explanation for the variable effects of lower doses is that CCK appears to interact synergistically with other eating-control signals, so that test conditions may be crucial. In addition, in both humans and rats, selective CCK-1 receptor antagonists have been shown to increase meal size (and the perception of hunger in humans) and to block the satiating effect of intraduodenal infusions of fat, in which CCK plays a significant role [56]. According to Geary's [21] scheme, CCK exemplifies a fully coupled endocrine satiating signal, i.e. the adequate stimulus (food in the small intestine) almost immediately leads to hormone secretion, which in turn affects eating within minutes. This tight linkage would seem to be an advantage both for the analysis of physiological mechanisms and for the development of pharmacotherapy. Whether long-term treatment with CCK or CCK agonists can be used effectively to control body weight, however, remains unclear [61, 62].

The effects of spontaneous mutations in the CCK-1 receptor to induce overeating and obesity lend further support to CCK's physiological role [21, 58]. The complication is that in rats and humans, CCK is also a CNS neurotransmitter, and CCK-1 receptors in the dorsomedial hypothalamus appear to mediate eating effects [63]. Thus, some of the phenotype of the knockout animals might be related to purely CNS CCK.

Intestinal CCK's satiating action appears to arise locally, in the gut. For example, Cox et al. [64] found that doses of CCK or of CCK-1R antagonists that had no effect on eating in rats when infused systemically were sufficient to affect eating when infused into the superior pancreatico-duodenal artery, which perfuses the pyloric area, the proximal duodenum and the pancreas. This local action of CCK appears to elicit a vagal afferent signal because subdiaphragmatic vagal deafferentiation (SDA) is sufficient to block the satiating effect of exogenous CCK. These and many studies of neural activation using c-Fos immunocytochemistry imply that the central neural processing of CCK satiation begins in the NTS. This is consistent with many subsequent findings, including some reviewed here (see chapters by Baskin and Blevins [65] and Schwartz [49]).

Intestinal Glucagon-Like Peptide 1 (GLP-1)

The active form of GLP-1 (fig. 5), GLP-1[7–36 amide], is synthesized by L-cells mainly in the jejunum and is released during and after meals, especially carbohydrate- or fat-containing meals. Evidence suggesting that GLP-1 elicits satiation and perhaps post-prandial satiety has accumulated rapidly in recent years. Other data suggest a similar role for peptide YY (PYY), which is released from the same L-cells [66–69]. Remotely controlled intraperitoneal or hepatic-portal infusions of GLP-1 during spontaneous meals selectively reduced meal size in rats [70], but whether physiological doses of GLP-1 were sufficient for these effects was not established. The situation in tests of humans is similar [24]. So far, administration of a GLP-1 antagonist has been reported to increase eating in rats in only one study, and then under rather limited conditions [71]. The conclusion of Williams et al. [71] was that endogenous GLP-1 is sometimes involved in the control of eating, but that the circumstances under which this happens and why the phenomenon is not more general, requires further work.

The study of GLP-1's physiological effects is complicated by the fact that it is rapidly broken down by the enzyme dipeptidyl-peptidase IV (DPP-IV), which is expressed in most capillaries, so that only a fraction of intestinal GLP-1 released during meals reaches the liver, and even less reaches the general circulation. For this reason, the GLP-1 analog exendin-4 (Ex-4), which is not rapidly cleaved by DPP-IV, is often used. Peripheral administration of Ex-4 produces a potent and lasting inhibition of eating [72, 73]. Administration of GLP-1 or of Ex-4 directly into the PVN or of Ex-4 into the dorsal hindbrain also inhibit eating [72, 74]. Ex-4, however, has biological potency orders of magnitude higher than that of GLP-1 [75], so studies using it require very cautious interpretation. In particular, it remains uncertain whether sufficient intestinal GLP-1 reaches the systemic circulation to affect posthepatic sites. An alternative hypothesis is that GLP-1 acts locally on vagal nerve endings in the lamina propria of the intestinal mucosa before entering the mesenteric capillaries [70].

We recently observed that the satiating action of intraperitonal infusions of GLP-1 during spontaneous meals was substantially reduced in rats with SDA, whereas the satiating action of hepatic-portal infusions of GLP-1 was not [70]. These data suggest that exogenous GLP-1 can act in more than one site to inhibit eating, that one of the sites is preferentially accessed by intraperitoneal infusions, and that GLP-1 acting at this latter site inhibits eating via a vagal afferent signal. Whether the same is true for endogenous GLP-1 remains to be determined.

Amylin

Four hormones produced by the pancreatic islets, insulin, glucagon, somatostatin and amylin, or islet amyloid polypeptide, have been implicated in the control of eating [76]. Of these, amylin is most actively investigated these days, both as an acute satiation signal, as described here, and as an adiposity signal, as described in the chapter by Lutz [77]. Amylin is synthesized by pancreatic beta cells and co-secreted with insulin beginning in the first minutes of meals. Intraperitonal injection of amylin just before meals or hepatic portal vein infusion of amylin during meals dose-dependently reduces meal size in rats [76, 78–80]. The smallest effective doses to inhibit eating were about double the endogenous levels [81], so whether amylin meets the physiological dose criterion is not certain. The failure of exogenous amylin to mimic the dynamics of endogenous secretion or, as discussed above, the lack of endogenous synergies may explain the apparent failure. More conclusively, the amylin receptor antagonist AC187 increased meal size in rats [82, 83]. Amylin's satiating effect has not been investigated in detail in humans. Amylin acts on receptors in the area postrema (AP) to inhibit eating. Lesion of the AP eliminates its effect, direct administration of amylin into the AP inhibits eating, and AP administration of AC187 increases eating [82].

Metabolic Signals in the Control of Eating

Introduction

Eating is part of the homeostatic regulation of body weight and of the availability of metabolites and essential nutrients. Physiological principles therefore suggest that metabolism feeds back to control eating. Parenteral administration of metabolic fuels often reduced food intake, whereas pharmacologic inhibition of fuel utilization increased it, and metabolic inhibitors also attenuated the eating-inhibitory effects of intravenous nutrient infusions [84]. This suggests that fluctuations in the availability or utilization of energy-yielding substrates – mainly glucose and fatty acids – or a common denominator of their utilization, control eating. Sensing of fuel availability or utilization leading to altered eating occurs in both the periphery and the brain [85, 86] (fig. 6). Unresolved is whether the effects of metabolic inhibitors are physiologically relevant or only emergency responses. While the threshold decrease in glucose utilization or fatty acid oxidation for a stimulation of eating is probably greater than what occurs before spontaneous meals, the fact that a signal is rarely activated in affluent people who eat three or more scheduled, ample meals each day does not necessarily mean that it is un-physiological. Also, if an integrated metabolic signal contributes to meal initiation, a pharmacological change in the utilization of a single metabolite might well be required to trigger a meal.

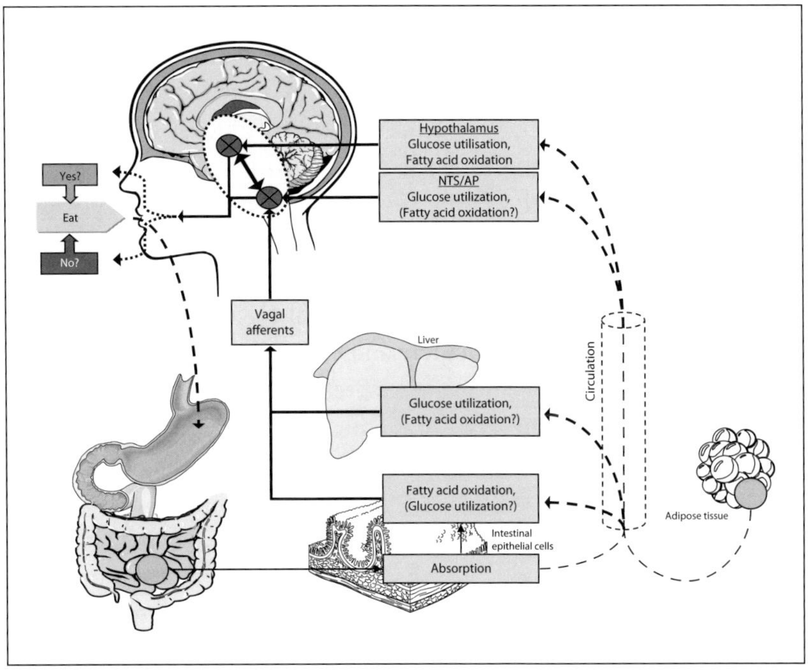

Fig. 6. Peripheral and central nervous system sensors that react to the availability or utilization of metabolic fuels affecting eating. Circulating metabolic substrates derived from absorption or from the mobilization of endogenous stores (i.e. glucose from the liver or free fatty acids from the adipose tissue) may reach the brain via the circulation or trigger vagal or other peripheral neural afferent signals. Signals reaching the brain may act in the caudal brainstem, especially the NTS and AP, or in the hypothalamus, especially the Arc. The bidirectional arrow between the hypothalamus and caudal brainstem indicates the important interconnections of these areas in translating feedback signals into altered eating behavior, as explained in the text. Modified with permission from Langhans et al. [281].

Signals Derived from Glucose

A small but consistent decline in blood glucose levels prior to spontaneous meals has been described in rats [87] and man [88] and may act as a pattern whose recognition contributes to meal initiation [89]. It is unclear which mechanism causes blood glucose to decrease prior to meals and whether this is accompanied by a decrease in glucose utilization. Blood glucose concentration and glucose utilization increase substantially in response to carbohydrate ingestion, and intravenous glucose infusions have often been shown to inhibit eating [84]. In some studies the satiating potency of glucose was increased by insulin [90], suggesting that the involved glucose sensors are partly sensitive to insulin. Studies in transgenic mice lacking the glucose transporter-2 (GLUT-2) [91] provide evidence for a physiological role of glucose in the control of eating: GLUT2-KO mice that express a transgenic glucose transporter only in their beta cells so as to rescue insulin secretion eat substantially more than corresponding wild-type (WT) mice and show increased hypothalamic orexigenic and decreased anorexigenic neuropeptide expression during the fasted-to-fed transition [91]. Thus, the absence of GLUT2 compromised the function of glucose sensors which are involved in the control of eating and influence hypothalamic neuropeptides.

Because of its unique location and function, the liver was considered likely to be involved in the control of food intake early on [92]. Infusion of physiologic amounts of glucose into the hepatic portal vein (HPV) reduces food intake more than equivalent infusions into the jugular vein [93–95], and intrameal HPV infusions of small amounts of glucose or glucose and insulin acutely and selectively reduced spontaneous meal size in the rat [96]. Thus, a meal-related increase in hepatic portal glucose concentration may contribute to satiation (fig. 6). The available electrophysiological and anatomical data indicate that vagal afferents terminating in the wall of the HPV function as hepatic glucose sensors, as originally suggested by Niijima [97].

In the brain, glucose-sensing neurons, i.e. neurons that regulate their membrane potential and firing rate in response to glucose, are present at different levels from the hindbrain to the hypothalamus (fig. 6) [98] and, together with peripheral glucose sensors, represent an anatomical and functional network that monitors glucose availability and is involved in glucose homeostasis and food intake control [99]. Glucose phosphorylation by glucokinase (GK) is the rate-limiting step in ATP production and is essential for effects of glucose on membrane potential and ion channel function of glucose-sensing neurons. GK, GLUT2, the sulfonylurea receptor-1 (SUR1), and the GLP-1 receptor are co-localized in several brain areas [100, 101] and have been proposed to be involved in central glucose sensing and control of food intake, but the exact role of GLUT2 in brain glucose-sensing is not fully understood [100, 102]. Glucose-sensing neurons also change their firing rate in response to other metabolites and hormones (e.g. insulin, leptin) [103], i.e. they appear to integrate different inputs, and their output controls neuroendocrine and autonomic responses as well as eating. Also, glucose availability influences the expression and turnover of several catabolic and anabolic neuropeptides [103] which presumably mediate the effects of glucose-sensing on eating. These hypothalamic circuits are discussed in detail in the chapter by Moran [86].

Signals Derived from Fatty Acids

Acute pharmacologic inhibition of fatty acid oxidation (FAO) is usually accompanied by a stimulation of eating in animals and man [104]. Some findings suggest that the current rate of FAO is crucial for this effect. In contrast, long-term inhibition of peripheral FAO by chronic administration of the carnitine palmitoyl-transferase (CPT-1) inhibitor etomoxir in rats increased muscle and liver fat content and induced insulin resistance, but did not induce hyperphagia [105]. Also, transgenic mice with reduced peripheral FAO and humans with genetic disturbances in fatty acid metabolism are not hyperphagic or obese [106, 107]. Together, these findings suggest that chronic inhibition of peripheral FAO does not affect eating.

The prominent role of FAO as an energy source for the liver suggested the hypothesis that hepatic FAO sensors generate signals that affect eating [84, 108]. Several recent findings, however, question the hypothesis that hepatic fatty acid oxidation influences eating and suggest that there is an alternative, or at least an additional, site where fatty acid oxidation is sensed [109]. Nevertheless, it is clear that the eating-stimulatory effect of intraperitoneal administration of the fatty acid oxidation inhibitor mercaptoacetate originates in the abdomen because it was completely blocked by subdiaphragmatic vagal deafferentiation [110]. Together these findings therefore suggest that MA acts in the intestine to stimulate eating. This idea and the more general possibility that enterocytes may act as energy flow sensors in the control of eating are discussed in more detail in the chapter by Langhans [85]. Finally, fatty acids and/or fatty acid metabolism can also be sensed centrally, in the mediobasal hypothalamus, and this also affects eating (fig. 6) [86, 111]. As discussed in the chapter by Moran [86], the physiological relevance of this effect is still unclear.

An Integrated Metabolic Signal

The recent identification of the molecular switches and signaling pathways in cellular metabolism has spurred a revival of old hypotheses proposing that eating is controlled by an integrated 'energostatic' or 'ischymetric' signal rather than by the utilization of one particular metabolite (see [84], for review). Reduction of cellular energy availability due to a decrease in fatty acid oxidation or glucose utilization increases the AMP/ATP ratio and activates the ubiquitous cellular energy sensor AMP kinase (AMPK) which exists in the periphery and the brain. AMPK activation or deactivation in the hypothalamus increases or decreases food intake [112, 113], suggesting that changes in cellular energy status contribute to the control of eating. The mammalian target of rapamycin (mTOR) is another cellular sensor of fuel availability and energy [114], and increased mTOR signaling in the hypothalamus decreased food intake and body weight in the rat. mTOR appears to colocalize mainly with Arc NPY/AgRP neurons [114]. Interestingly, central administration of L-leucine also increased hypothalamic mTOR and decreased food intake and body weight. As mTOR stimulates protein synthesis, these findings suggest that mTOR is involved in the control of cell growth and proliferation by energy availability. AMPK and mTOR both also respond to hormones involved in the control of energy balance (AMPK to leptin and ghrelin, mTOR to leptin) and thus may represent cellular sensors that integrate fuel availability and endocrine signals. In contrast to mTOR, AMPK activity is increased by fuel deficiency and decreased by metabolites and leptin [113], and activation of AMPK inhibits mTOR activity [114], suggesting that these fuel-sensitive kinases have reciprocal functions. An emerging concept is that changes in AMPK- and mTOR-sensing in the brain in response to fuel surplus inhibit eating, whereas similar changes in the periphery may limit nutrient uptake into tissues, i.e. cause insulin resistance.

Adiposity Signals

Introduction

As described above, there is evidence for an active physiological regulation of long-term energy balance and, therefore, body weight in adults. Perhaps the strongest evidence for such regulation are the many reports that, in both animals and humans, experimental manipulations of body weight in adults provoke compensatory changes in energy intake and expenditure that serve to return weight to the normal level (see [115, 116], for reviews).

Given the obvious epidemiological evidence that western populations are rapidly growing markedly fatter, however, it is equally clear that this regulatory system does not work perfectly in the environment in which most of us live. Nevertheless, the fact that even small constant errors in the balance between energy intake and expenditure would lead to much larger body weight gains than we are actually experiencing suggests that the regulatory system is actually quite powerful – for example, a constant positive imbalance of only 1% would lead to a gain of over 1 kg/year adipose tissue. Although most humans gain weight during the decades of middle age, very few gain the more than 30 kg that this calculation suggests.

The fact that body weight changes in adult individuals are mainly due to fluctuations in body adiposity suggests that the level or state of adiposity is the regulated variable. What aspect of adiposity does the brain sense? Over 50 years ago, Kennedy [117] hypothesized that circulating factors whose plasma levels reflect the size of the fat stores regulate adiposity by controlling food intake and energy expenditure. These signals were originally called *lipostatic signals*; these days the term *adiposity signals* is favored. The basal levels of leptin, insulin, amylin and other hormones may function as such signals (fig. 7). The following paragraphs review some of the principal evidence in favor of leptin and insulin, and several chapters take up current issues related to these candidate adiposity signals [5, 49, 65, 77].

Leptin

A series of elegant experiments demonstrated in the 1970s that the dramatic obesity and diabetes phenotypes of *ob/ob* and *db/db* mice were caused by single-gene mutations of an unknown hormone and its receptor, respectively [118, 119]. A significant new chapter in the physiology of eating opened in 1994 when Zhang et al. [120] used molecular genetic methods to identify the adipocyte hormone leptin

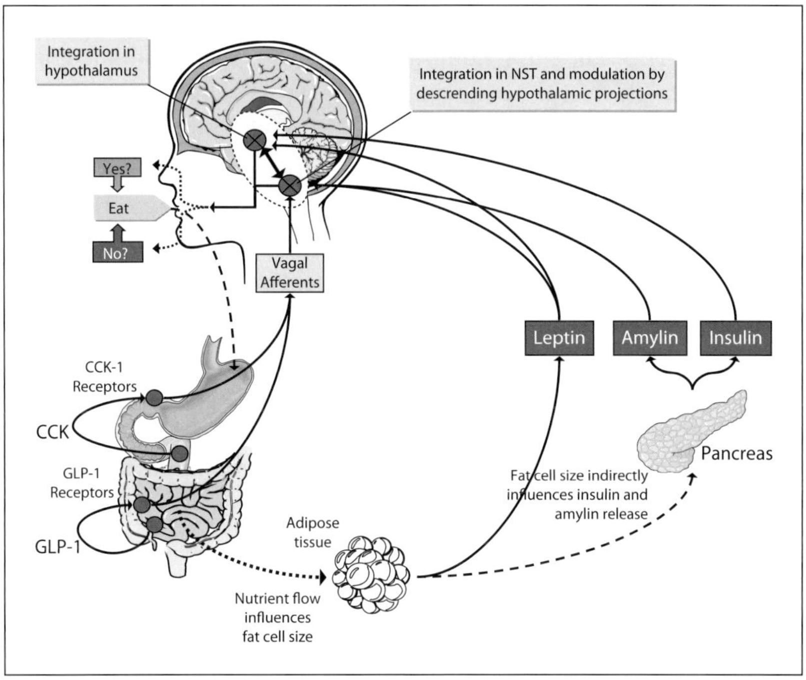

Fig. 7. Adiposity signals (leptin, insulin and amylin) affect eating by modulating the action of meal-related, mainly vagally mediated satiation signals, such as cholecystokinin (CCK) and glucagon-like peptide-1 (GLP-1). Leptin may act on receptor both in the caudal brainstem and hypothalamus; insulin acts in the hypothalamaus; and amylin acts in the AP. The hypothalamic actions of leptin and (presumably) insulin activate descending pathways to the caudal hindbrain. See text for further details.

as the missing signal in *ob/ob* mice. This was quickly followed by identification of the leptin receptor and its *db* mutation [121, 122]. The human leptin gene is now known as *LEP*, its mouse homolog as *lep*, and these mutations as lep^{ob} and LR^{db}. Six variants of the leptin receptor, LR, have been discovered in mice; the long, signaling form is LRb.

Several lines of evidence beyond these gene mutation syndromes support the role of leptin as an adiposity signal (fig. 7). Cross-sectional studies have revealed high correlations between basal leptin levels and adiposity in humans and animals [5]. Leptin is actively transported into the Arc and binds to LRb, on two populations of Arc neurons that contribute to the control of eating (see below), and local injections of leptin into this area or the adjacent third cerebral ventricle reduce food intake, increase energy expenditure, and reduce body weight in rats and mice [123–125]. LRb are located in other brain areas as well, and local administration of leptin in these areas also reduces food intake [126–128]. Peripheral administration of leptin also reduces food intake, by selectively decreasing meal size [129, 130].

Perhaps the strongest physiological evidence that leptin is an adiposity signal is the report by Zhang et al. [131] in 2007 that continuous infusion of a leptin antagonist into the third cerebral ventricle over the course of several days led to increased eating and body weight. These data strongly implicate leptin in the physiological control of eating, although they do not directly link leptin to adiposity signaling. Tests in overweight human subjects who lost weight by dieting have produced evidence that leptin meets the physiological dose criterion for an adiposity signal, at least during underweight [132]. After the subjects lost weight, their basal energy expenditure decreased (eating was not measured). Then, leptin was infused in amounts that re-established pre-dieting leptin levels. This was sufficient to return basal energy expenditure to the pre-dieting level. This interesting result is one of several that supports the hypothesis the reduced plasma leptin levels affect eating and energy expenditure more potently than do increased plasma levels, suggesting that leptin may function physiologically as a starvation signal more than as an obesity signal [21, 131–134].

Insulin

Basal plasma and cerebrospinal levels of insulin are equally tightly linked to body adiposity, insulin receptors are present in the hypothalamus, and the actions of central insulin on food intake and energy expenditure are similar to those of leptin in many respects [5, 135, 136] (fig. 7). Moreover, male and female mice with genetic deletions of neuronal insulin receptors are obese and female mice are also hyperphagic [137], indicating that insulin receptor signaling in the brain is important for the control of body weight. Insulin crosses the BBB via a receptor-mediated process [138], and it acts through the same hypothalamic neuropeptide system as leptin [139].

From Long-Term Energy Balance to Single Meals

Any signal which controls body weight by changing food intake must modulate the frequency or the size of single meals and, therefore, must interact with the shorter-term, meal-control signals. As described above, exogenous leptin and insulin selectively reduce meal size [129, 130, 140, 141], so should interact with reward or satiation signals, which also affect meal size. In line with this, both leptin [142] and insulin [143] have been shown to enhance the satiating effect of CCK, although the insulin effect may not be a selective meal size effect. Finally, the compensatory hypophagia that follows experimentally induced increases in body weight is also mainly due to a reduction of nocturnal meal size, further supporting the hypothesis that adiposity signals influence eating mainly through changes in meal size [116]. In their chapters, Blevins and Baskin [65] and Schwartz [49] describe recent progress on the mechanisms underlying this interaction.

Central Nervous System Integration

Introduction

Eating is mediated by a very complex, anatomically diffuse neural network that is organized hierarchically, redundantly and recurrently. This section introduces some principal nodes in this network, their key signaling molecules, and their main functions in the control of eating and regulation of body weight, as presently understood. Both the discovery of new facts and the generation of new concepts are proceeding rapidly in this area, as reflected in the chapters by Blevins and Baskin [65], Bouret [144], Sullivan and Grove [145], and Schwartz [49]. To introduce these developments, we begin with a historical perspective, aimed at providing a sense of the evolution of how the CNS mechanisms controlling eating have been analyzed and interpreted. We then discuss some of the key anatomical nodes and neurochemical signaling molecules. For reasons that will become clear in the next section, we begin with the hypothalamus.

The experimental analysis of the integrative action of the CNS in the control of eating has progressed in overlapping waves, each initiated by methodological advances. The first wave began six decades ago with the development of stereotaxic surgery. This method led to the discoveries that circumscribed lesions of the ventromedial hypothalamic area (VMH) induce hyperphagia, reductions in energy expenditure, and weight gain and that similar lesions of the lateral hypothalamic area (LHA) induce opposite effects [146, 147]. This work led directly to the concept of hypothalamic 'centers' for eating and weight regulation (fig. 8) [148]. During the subsequent decades, lesion and neuropharmacological work elaborated and better differentiated the functions of these areas [149–151]. Also, the Arc, paraventricular (PVN) and dorsomedial (DMN) hypothalamic nuclei as well as several nonhypothalamic areas were implicated in the neural circuitry for eating and weight regulation [152–154].

A second wave began around 1970, with the advances in neuroanatomical methods, especially fluorescence, immunocytochemical and tract-tracing methods. These led to a new, chemical neuroanatomy [155–158]. Early landmarks in this era include the demonstrations that adrenergic receptors in part mediate the hypothalamic control of eating [159], that chemical lesions of ascending dopaminergic pathways traversing the LHA are sufficient to replicate the syndrome of aphagia and adipsia produced by electrolytic lesions of the LHA [160], and that descending oxytocin projections from the hypothalamus to the caudal brainstem contribute to hypothalamic lesion-induced obesity [161]. An especially important development was the increasing realization that eating and related neuroendocrine and autonomic responses are coordinately organized by a diffuse neural network extending from the cerebral cortex and basal telencephalic structures caudally through the hypothalamus and into the caudal brainstem [162, 163]. As a consequence, eating and weight-regulatory functions cannot be localized to particular discrete 'centers'. Therefore, we do not use this terminology here.

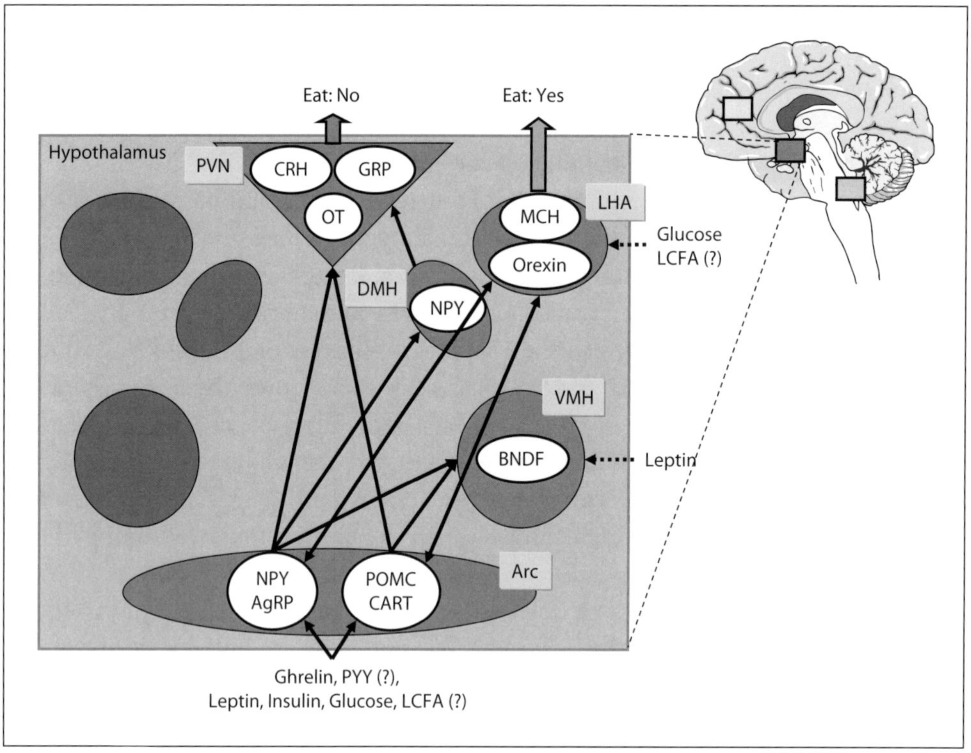

Fig. 8. Schematic frontal section of the hypothalamus indicating the localization of cell bodies expressing orexigenic and anorexigenic neuropeptides, some of their intra-hypothalamic projections, their hormone and metabolite sensitivities and the putative functional roles of extrahypothalamic projections form the PVN and LHA (based on the effects of orexigenic and anorexigenic neuropeptide administration). Note the bilateral symmetry of the hypothalamus (labels and projections are shown only unilaterally). Hypothalamic areas: Arc = Arcuate nucleus; LHA = lateral hypothalamic area; PVN = paraventricular nucleus; VMH = ventromedial hypothalamic area. Neuropeptides: AgRP = agouti-related peptide; BDNF = brain-derived neurotropic factor; CART = cocaine- and amphetamine-related transcript; CRH = corticotropin-releasing hormone; GRP = gastrin-releasing peptide; MCH = melanin-concentrating hormone; NPY = neuropeptide Y; OT = oxytocin; POMC = pro-opiomelanocortin. Metabolites: LCFA = long-chain fatty acids. See text for further details. Modified with permission from Langhans et al. [280].

A third wave of progress, based on the application of molecular genetic techniques, began, as described above, with the discovery of leptin [120] in 1994 and the leptin receptor [121] within 2 years. The first demonstrations that exogenous leptin acts in the brain to inhibit eating and to restrain adiposity came in 1995 [123–125]. The outlines of the CNS mechanisms for this effect emerged soon after. By 1996, it had been shown that leptin crosses the BBB via a saturable carrier system which is especially active in the Arc [164], that Arc neurons densely express LRb mRNA [165], that these same neurons also express mRNA for the neurotransmitter neuropeptide Y (NPY) [166], whose administration stimulates eating, and for pro-opiomelanocortin (POMC), from which the eating-inhibitory neurotransmitter α-melanocyte-stimulating hormone (α-MSH) is cleaved, and that antagonism of hypothalamic α-MSH signaling via melanocortin-4 receptors (MC4R) blocks the eating-inhibitory effect of exogenous leptin [167]. We now know that these same neurons also function as metabolic sensors and express receptors for insulin, ghrelin and serotonin, to name just a few of the substances emphasized here. These studies provided a novel window into the brain networks controlling eating, whose elaboration continues [114, 168–173].

A fourth wave has already begun, based on increasingly sophisticated functional imaging methods [6, 174–176]. Because many of these methods can be linked explicitly to molecular and genetic techniques, they hold unprecedented promise for translating basic animal research into human functional neuroscience and into therapeutics. The chapters by Neary and Batterham [2], Kringelbach and Stein [3], and Stice and Dagher [4] describe some of these developments.

Hypothalamus

Arcuate nucleus (Arc). Figure 9 shows a schematic of some critical aspects of the organization of the Arc [114, 168–173]. A key role in weight regulation is suggested by the fact that, as described above, receptors for the putative adiposity signals leptin, insulin, and ghrelin are expressed by Arc NPY and POMC (α-MSH) neurons. The former also expresses agouti-related peptide (AgRP), another eating-stimulatory peptide, and the latter, cocaine- and amphetamine-related transcript (CART), which inhibits eating. Several mechanisms contribute to the functional coordination of these two sets of neurons. For example, the activity of the POMC/CART neurons is inhibited by NPY acting at Y1 receptors on POMC cell bodies [177, 178] and by gamma-amino butyric acid (GABA)-mediated inhibitory synapses on projections from the NPY/AgRP neurons. Both populations of Arc neurons also receive serotoninergic (5HT) inputs, the POMC/CART neurons via 5HT-2CR and the NPY/AgRP neurons via 5HT-1BR, which modulate their activity in a similar fashion as leptin [172]. Finally, the principal projection targets of these Arc neurons are the PVN and LH. This basic circuit is thought to orchestrate the eating, neuroendocrine and autonomic responses contributing to energy homeostasis. In particular, increases in adiposity are thought to generate catabolic responses (i.e. the inhibition of eating and stimulation of energy expenditure) through this circuit, and decreases in adiposity to generate anabolic responses (stimulation of eating and inhibition of energy expenditure). Chronic pharmacological stimulation of the Arc-PVN NPY system stimulates eating, reduces energy expenditure, and results in obesity [179–181]. In contrast, chronic administration of NPY antisense mRNA into the Arc decreased food intake and reduced body weight [38]. Both intracerebroventricular insulin and intracerebroventricular leptin reduce Arc NPY mRNA. Furthermore, NPY neuronal activity is increased in animals in which body weight has been reduced by food restriction and is decreased in dietary-obese animals [182]. In

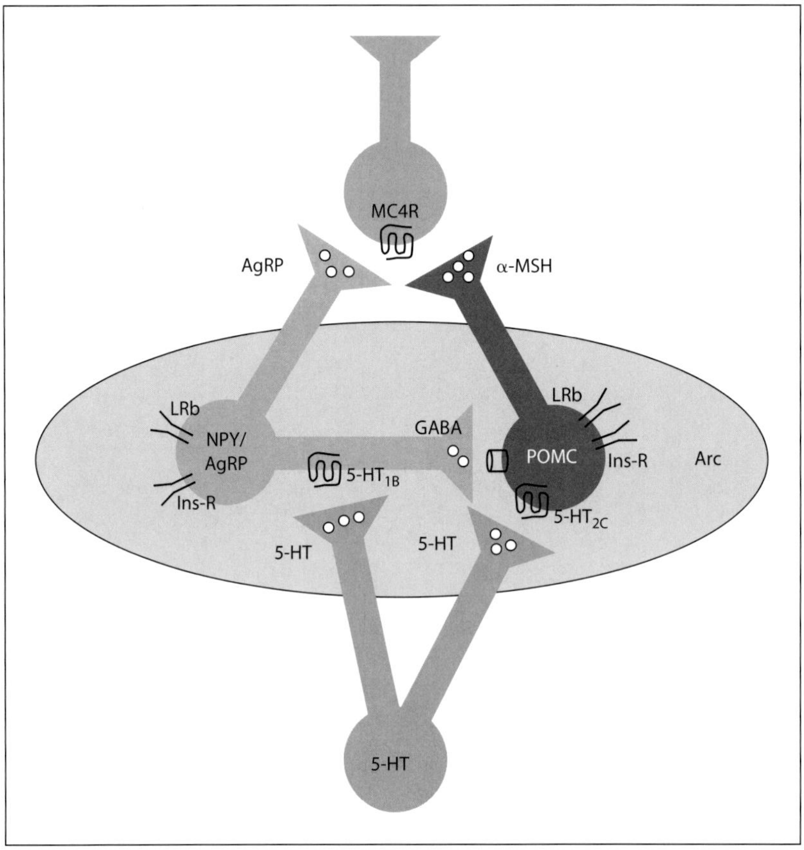

Fig. 9. Schematic of the current concept of the connectivity of NPY/AgRP and POMC neurons in the Arc that control eating via MC4R neurons in the PVN and other hypothalamic sites. 5-HT neurons reach the Arc from the Raphe nuclei in the brainstem. 5-HT = Serotonin; 5-HT_{1B} and 5-HT_{2C}, 1B and 2C subtype serotonin receptors; AgRP = agouti-related peptide; α-MSH = alpha-melanocyte-stimulating hormone; GABA = gamma-amino butyric acid; Ins-R = insulin receptor; LRb = signaling form of the leptin receptor MC4R, melanocortin-4 receptor; NPY = neuropeptide Y; POMC = pro-opio-melanocortin. Neurons expressing α-MSH. See text for further details. Modified with permission from Langhans et al. [280].

contrast to MC4R KO mice, however, NPY or NPY receptor KO mice eat normally, presumably due to redundancy and developmental compensation. This is an example of the caution required in drawing physiological conclusions from knockout phenotypes.

The synthesis presented above is based on a tremendous data base. Nevertheless, because available methodologies do not permit truly crucial physiological experiments, the extent to which it reflects normal physiological function is uncertain. Some outstanding issues that are currently under investigation include: (1) the extent to which adiposity signals act at sites outside the Arc, for example, the caudal brainstem (fig. 10) [162], (2) whether Arc signaling is more physiologically active during

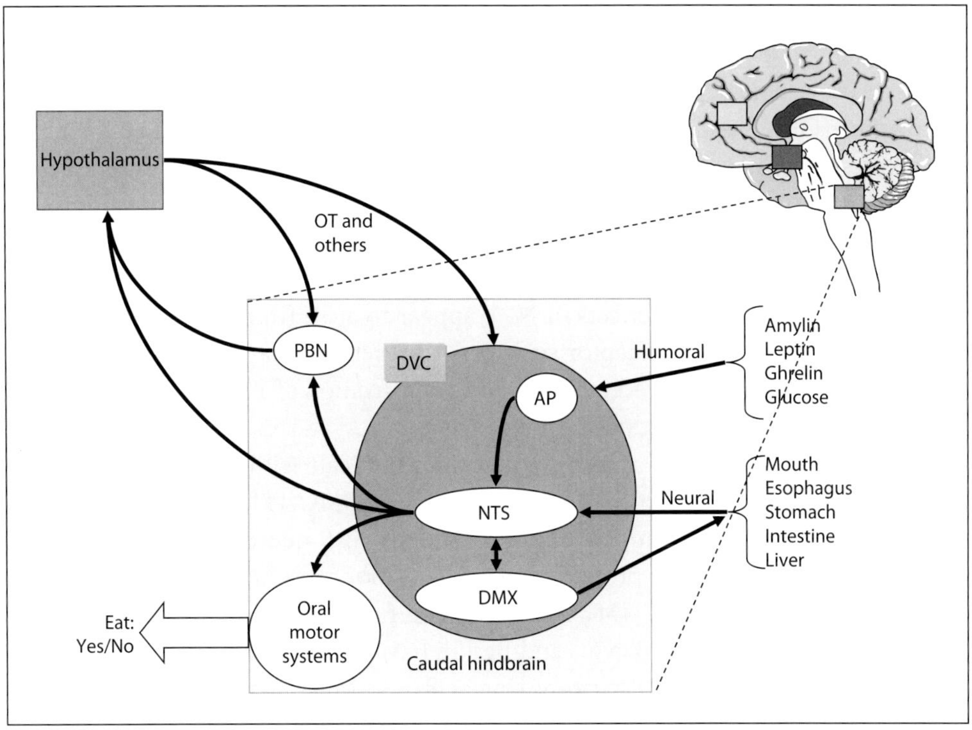

Fig. 10. Schematic of some principal caudal brainstem areas implicated in the control of eating, their neural connections, and their peripheral sensitivities. AP = Area postrema; DMX = dorsal motor nucleus of the vagus; DVC = dorsal vagal complex; NTS = nucleus tractus solitarii; OT = oxytocin; PBN = parabrachial nucleus. See text for further details. Reproduced with permission from Langhans et al. [280].

underweight or during overweight, (3) what the physiologically relevant dynamic properties of the circuit are, for example, how do the state of adiposity and the current flux of energy substrate combine to affect its activity, (4) whether the effects on eating and energy expenditure are always coordinated, and what factors might dissociate them [183], and (5) how lasting changes in adiposity affect the operation of the system. The last question is especially interesting with regard to the well-recognized progressive resistance of obese subjects to repeated doses of leptin [165, 184] or insulin [76] (see chapters by Banks [20] and Münzberg [185]).

Paraventricular Nucleus (PVN) and Lateral Hypothalamic Area (LHA). Figure 8 depicts the main Arc projections controlling eating. In the PVN, α-MSH terminals synapse on MC3R and MC4R [186–188], and NPY terminals synapse on Y1, Y4 and Y5 NPY receptor subtypes [179, 189]. PVN neurons express several anorexigenic neuroactive substances, including corticotropin-releasing hormone (CRH), oxytocin (OT), gastric-releasing peptide (GRP), and thyrotropin-releasing hormone (TRH). For this reason, and because PVN lesions produce hyperphagia and obesity, the PVN seems to be predominately a catabolic integratory site. In contrast, the LHA appears to be a mainly anabolic site. It receives many orexigenic projections from the Arc, expresses orexigenic peptides including melanin-concentrating hormone (MCH) and the orexins (ORX), and LHA lesions produce aphagia. Nevertheless, LHA neurons also express anorexigenic substances, such as CART and dynorphin. Interestingly, in contrast to the more unidirectional Arc-PVN connections, there are prominent reciprocal LHA-Arc projections, which may be related to the more heterogeneous expression profiles of its neurons.

Although the orexigenic effects of NPY appear to arise from a synergistic activation of Y1, Y2, Y4 and Y5 receptor subtypes, YI receptors may be especially important. Genetic deletion of Y1 receptors or the administration of Y1 receptor antagonists reduces NPY- and fasting-induced eating [190–194].

One function of the PVN apparently is to communicate with the caudal hindbrain areas involved in the control of eating. For example, POMC, OT and CRH neurons in the parvocellular subdivision of the PVN as well as GRP neurons in the magnocelluar subdivision of the PVN project to the nucleus tractus solitarii (NTS) and dorsal motor nucleus of the vagus [158, 195–198]. The functions of these projections are discussed in the chapter by Blevins and Baskin [65] and Schwartz [49].

In the LHA, MCH- and ORX-expressing neurons receive synaptic inputs from Arc NPY-, AgRP- and α-MSH-expressing neurons [199–201]. Chronic central administration of MCH results in increased food intake and adiposity [202], whereas chronic administration of MCH-1 receptor antagonists inhibit eating and reduce body weight [203]. Additionally, transgenic mice overexpressing MCH are hyperphagic and obese, whereas MCH-null mice are hypophagic and lean [204]. MCH affects energy expenditure as well as eating. ORX A and B are 33 and 28 amino acid peptides, respectively, that increase arousal and stimulate eating [205, 206]. Finally, as discussed previously, the LHA also contains neurons that function as receptors for glucose and, perhaps, long chain fatty acids (fig. 8) [207, 208]. ORX may play a role in glucose-sensing because hypoglycemia induces increases in ORX mRNA and c-Fos expression in ORX neurons [209], and ORX-A excites LHA glucose-sensing neurons [210, 211]. Finally, LHA ORX neurons reciprocally innervate NPY- and POMC-producing neurons of the Arc [211].

Other Hypothalamic Areas. As suggested by figure 8, the Arc, PVN and LHA are by no means the only hypothalamic areas contributing to the control of eating. For example, direct injections of NPY into the perifornical area and ventromedial hypothalamus (VMH) stimulate eating similarly to PVN injections [179]. As well, NPY projections to the PVN from sites other than the Arc, such as the dorsomedial hypothalamus (DMH) have also been implicated in the control of eating. This DMH projection is especially interesting. DMH NPY neurons are under a tonic inhibitory influence of neuronal (not hormonal) CCK, such that rats with mutations of the CCK-1 receptor overexpress NPY in the DMH and are obese [63]. In addition, chronic increases in exercise seem to produce an independent tonic inhibition of DMH NPY [212].

The VMH also receives projections from Arc NPY/AgRP and POMC neurons, so has been regarded as another downstream site for signaling events initiated in the Arc. The VMH also contains LRb, so is a potential sensory site for adiposity signals. Dhillon et al. [213] recently provided support for this idea by demonstrating that leptin depolarizes and increases the firing rate of steroidogenic factor-1 (SF1)-positive neurons in the VMH and that transgenic mice lacking LRb on SF1-positive neurons became heavier than controls when fed an energy-dense diet. Brain-derived neurotrophic factor (BDNF) is another candidate mediator of the VMH's effect on in eating. BDNF is abundantly expressed within the VMH, and mice with reductions in BDNF neuronal function increase food intake and body weight [214].

Caudal Brainstem

Eating is a rhythmic behavior produced by motor neurons of cranial nerves V, VII, IX, X and XII. These nerves are driven by central pattern generators, a variety of reflex-like sensory feedback signals, and more remote upstream neural networks, including the hypothalalmic projections described above. The neural machinery for central pattern generators and sensory feedback signals are contained in the caudal brainstem. The caudal brainstem also receives the sensory inputs from all but the olfactory dimension of flavor and from a variety of interoceptive information. It also contains many of the same neuronal signaling sensitivities as found in the hypothalamus, including leptin, ghrelin, amylin, NPY, MC and estrogen receptors as well as POMC neurons [162, 168, 188, 215–217]. Some of this is depicted in figure 10.

Studies by Harvey Grill and his colleagues [127, 162, 218] of the chronic decerebrate (CD) rat, i.e. animals with midcollicular transections of the neuroaxis, and of direct administration of neuroactive substances into the caudal brainstem, indicate that the caudal brainstem is sufficient for nearly normal effects of taste and gastrointestinal feedback signals on meal size. For example, although CD rats do not initiate meals unless food is placed into the mouth, when this is done by intraoral infusion, they take well-defined meals terminated by passive refusal of more food. Furthermore, CD rats increase or decrease meal size in a normal way when sucrose concentration is varied [219] and have apparently normal sensitivity to peripheral CCK injection [220]. Although CD rats eat less after insulin or leptin administration [162], the caudal brainstem does not seem sufficient for regulation of energy homeostasis. That is because they do not increase meal size normally after food deprivation or when the number of daily opportunities to eat is reduced [221].

The structure in the caudal brainstem most investigated in relation to the control of eating is the NTS (fig. 10). It receives a wide variety of sensory information, has important integratory functions, and is a source of ascending projections to further integratory sites as well as descending projections that control behavioral and autonomic responses. Flavor-related information reaches the NTS directly via cranial nerves V, VII, IX and X. Cranial nerves VII (the facial nerve), IX (glossopharyngeal) and X (vagus) innervate taste buds and mediate the primary tastes (sweet, bitter, sour, salt, protein or umami, and, in rats, apparently starch taste). Receptors for temperature, mechanical stimulation, and certain chemicals, such as capsaicin (chili) convey flavor information via cranial nerve V (trigeminal), which synapses in the sensory trigeminal nucleus prior to projecting to the NTS. In addition, a variety of interoceptive information reaches the caudal brainstem in part directly, via the transport of certain metabolites (glucose) and hormones (leptin), and indirectly, via neural projections from either the adjacent AP, which has a porous BBB, or from the gut. As described above, vagal and spinal visceral sensory nerves relay temperature, mechanical, osmotic and chemical (metabolite and hormone concentrations) information from the gut to the NTS.

The NTS is an integratory site, not a mere relay. For example, the electrophysiological responses of second order taste neurons are affected by a range of eating-related information, including plasma glucose levels and preference and aversion learning [222]. Immunochemical detection of the expression of c-Fos protein, a marker of neuronal activation, reveals an even wider range of integratory effects. For example, the increase in the satiating potency of CCK by estradiol described below is associated with an increase in CCK-induced c-Fos expression in NTS neurons expressing the estradiol receptor-alpha (ERα), strongly suggesting that the interaction arises in the NTS [223, 224]. Similarly, the functionally synergistic inhibition of eating produced by co-administration of leptin and CCK or CCK1R antagonists is mirrored by a similar increase in the number of NTS cells expressing c-Fos [142, 225]. This integrative function is the focus of the chapter by Schwartz [49].

Forebrain

The increased complexity of the forebrain, or telencephalon, is a hallmark of human evolution. As with other categories of information, eating-related information is represented and re-represented in telencephalic areas. Not surprisingly, therefore, the telencephalic contributions to eating-related associations, cognitions, emotions, and motives, both conscious and unconscious, are very poorly understood [168]. Nevertheless, substantial progress has been made in studies of the telencephalic contributions to some aspects of food reward.

As described above, orosensory pleasure is a powerful controller of eating. Analyses of the telencephalic contribution to this reward function support several generalizations: (1) Many reward-related behaviors are similar in animals and humans [8, 226]. (2) Partially homologous neural mechanisms mediate food reward in animals and humans [9, 227, 228]. (3) The neural networks mediating food reward and those mediating other natural (e.g. sex, water when thirsty) and unnatural (e.g. drugs of abuse) rewards overlap very heavily. (4) Neural networks mediating food

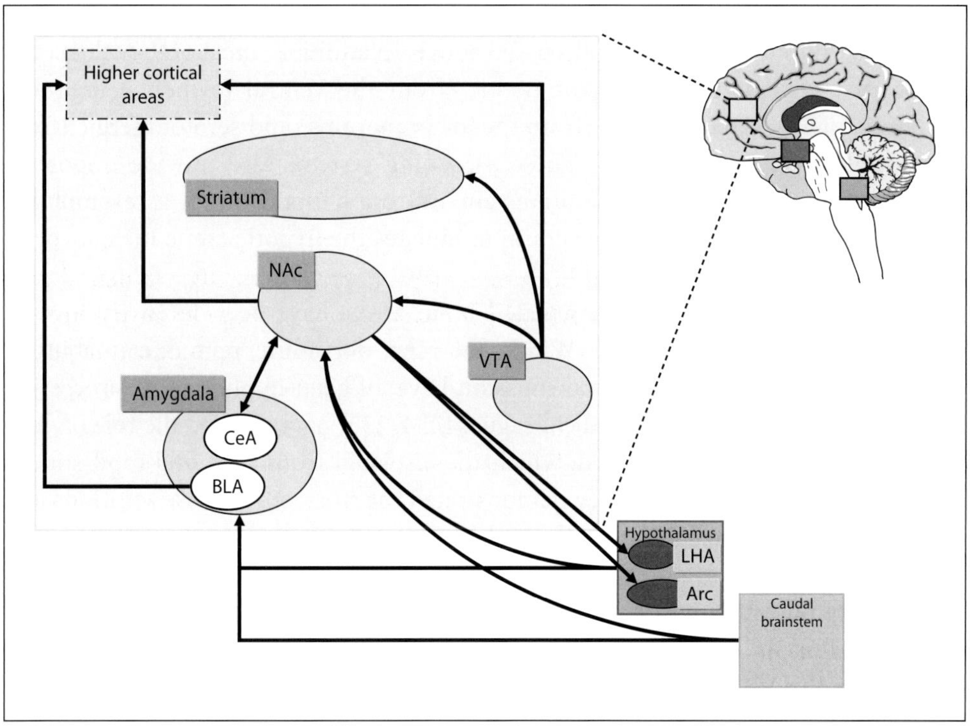

Fig. 11. Schematic of some principal forebrain areas implicated in the mediation of food reward and their neural connections. BLA = Basolateral nucleus of the amygdala; CeA = central nucleus of the amygdala; LHA = lateral hypothalamic area; NAc = nucleus accumbens; VTA = ventral tegmental area. See text for further details. Reproduced with permission from Langhans et al. [280].

reward also overlap with the mechanisms regulating mood and affect. (5) These neural networks are extensively and reciprocally connected with the neural networks mediating the more regulatory and reflexive aspects of eating, discussed above, so that simple notions of parallel processing or 'homeostatic vs. non-homeostatic' controls are correct in only the most general, heuristic way. An example of this is that ventral forebrain manipulations that are linked to hedonic eating also activate Arc NPY neurons and inhibit Arc POMC/CART neurons, whose activity is, as explained above, usually interpreted in the context of homeostasis [229]. (6) At least partially independent neural substrates can be identified for different aspects of food reward [226, 230–233]. The outstanding example of the last point is the differentiation of the affective or emotional impetus to eat (implicit 'liking' and conscious pleasure) from the classical motivational impetus (implicit incentive salience 'wanting' and cognitive incentive goals [9, 227].

Some principal components of the telencephalic reward system are the nucleus accumbens (NAc), the amygala, especially the central nucleus of the amygdala (CeA), and parts of the limbic, orbitofrontal, cingulate and insular cortical areas (fig. 11).

The NAc and the CeA, in particular, receive projections from a variety of hypothalamic areas and brainstem areas discussed above. In addition, most of the telencephalic reward system receives dopaminergic (from the ventral tegmental area and substantia nigra), noradrenergic (from the locus coerulus) and serotoninergic (from the rostral raphe nuclei) inputs. These ascending systems also provide important links among these areas and the brainstem and the hypothalamus, as exemplified by the recent report [172] that serotonin modulates the hypothalamic melanocortin pathway.

The NAc, and directly interconnected brain areas, have been intensely investigated in relation to food reward. Within the NAc, dopamine, opioid, cannabinoid, acetylcholine and GABA neurotransmission have all been implicated in processing food reward. Hajnal, Norgren and colleagues [234, 235] have verified the role of dopamine in orosensory food reward, which they isolated from post-oral food stimuli by testing sham-feeding rats. They demonstrated that for both sucrose solutions and corn oil emulsions, concentration-dependent increases in sham feeding were closely associated by the release of dopamine in the NAc. Note that these studies suggest that sensory information for two entirely different sensory pathways, i.e. relatively purely gustatory in the case of sucrose versus olfactory/trigeminal in the case of oil, converge in the NAc. Similarly, administration of mu-opioid agonists into the NAc preferentially stimulates ingestion of high-fat foods and sucrose solutions, whereas administration of opioid antagonists selectively reduces ingestion of palatable foods [236, 237]. Additionally, in man, opiate antagonists reduce food palatability, but not subjective hunger [238].

The endocannabinoid system also contributes to food reward circuitry. Endocannabinoids act at brain CB1 cannabinoid receptors in both the NAc and hypothalamus to stimulate eating, and endocannabinoid activity in these regions varies in relation to nutritional status and eating expression [239]. The powerful effects of manipulation of endocannabinoid function on mood, at least in susceptible individuals, exemplifies the overlap between neural mechanisms processing food reward and regulating emotion and affect [228].

Some aspects of the connectivity of the NAc, the amygdala, and other areas that mediate eating have emerged (fig. 11). For example, connections between the basolateral amygdala (BLA) and forebrain cortical regions appear crucial in determining food palatability [240]. A reciprocal connection between the CeA and the NAc is also involved in opioid-mediated eating [241], although whether this pathway is selectively involved in reward is not yet clear [242]. Another mechanism that may involve reciprocal projections from the NAc and ventral palladium to the LHA also appears to selectively stimulate consumption of palatable food to the LHA [243]. As previously discussed, the LHA contains MCH and orexin neurons, both of which stimulate food intake. Activation of opiate neurons in the NAc may stimulate eating by releasing these neurons from a tonic inhibition [244]. Similarly, the stimulation of eating caused by NAc administration of the GABA(A) agonist muscimol was associated with increased activity in orexin neurons in the perifornical region of the hypothalamus [229].

Physiological Modulators of Eating and Body Weight

A number of physiological functions whose primary purpose is not the control of eating nevertheless powerfully affect eating. For example, thermoregulation, water balance, exercise, and stress (both physiological and psychological) can affect eating in the short term and, under certain circumstances, chronically affect eating and weight regulation in the long term. This section discusses two such physiological modulators.

Sex

In the physiology of eating and body weight regulation, as in other areas, sex is a fundamental biological variable [245]. In animals and humans, reproductive, or hypothalamic-pituitary-gonadal (HPG), axis function affects the controls of eating, growth, energy metabolism, nutrient partitioning, physical activity, adipose tissue distribution, etc. Research in these areas is complicated by several factors, ranging from the extensive developmental and species differences in HPG axis function to the marked interactions of culture and physiology in most human behaviors related to HPG axis function.

After puberty, most physiological sex differences are not directly controlled by sex chromosomes, but by activational effects of the gonadal steroid hormones, i.e. effects related to current circulating levels of androgens, estrogens, and progestins. The clearest activational effect on eating is the decrease in eating that occurs during the peri-ovulatory period of the ovarian cycle in women and animals. This is absent when estradiol secretion does not occur, and a physiological pattern of estradiol treatment is sufficient to reinstate it in ovariectomized rats [246]. Part of the mechanism involves an increase in the satiating potency of CCK mediated by ERα stimulation in the NTS [223, 224, 247, 248]. Estradiol also appears to reduce the eating-stimulatory action of ghrelin [249]. Brain serotonin appears to be crucially involved in the effects of estradiol on eating [250]. Estrogens also exert activational effects on energy homeostasis and regional adipose tissue distribution, at least in part via ERα stimulation in the hypothalamus [251–253]. Emerging data also link membrane ERs to energy homeostasis, although the physiological relevance of these effects is still unclear [254]. Finally, the relevance of these controls of eating to human weight regulation is underscored by reports that loss-of-function polymorphisms of the ERα gene are linked to increases in fat mass in girls [255]. Several recent reviews [247, 248, 252, 256, 257] discuss the current progress in this important area.

The Immune System and Eating

The loss of appetite during illness is a well-known phenomenon resulting from the effects of immune activation on eating. Acute infections and other immune challenges trigger a generalized host defense reaction, known as the acute-phase response (APR), which is comprised of several physiological and behavioral changes, including anorexia [258]. Illness anorexia appears to be beneficial for the host in the beginning [259], but becomes deleterious over time. Most current knowledge on the mechanisms of illness anorexia is derived from the model of acute microbial infections produced by administration of lipopolysaccharides (LPS), Gram-negative bacterial cell wall constituents that are released in natural infections during bacteriolysis or rapid bacterial proliferation [260]. LPS administration stimulates the immune system and mimics the APR including anorexia. It initiates a cascade of immune and neuroendocrine events that involve endogenous mediators, such as pro-inflammatory cytokines, most likely acting at the BBB, where they trigger from BBB endothelial cells the release of other downstream mediators, such as prostaglandin E_2 or nitric oxide, to ultimately modulate the activity of the neuronal network described in the previous sections that controls normal eating [261]. A more detailed appraisal of the recent developments concerning this aspect of the control of eating is given in several recent reviews [261–263]. Because immune mechanisms are also implicated in obesity and diabetes (see below), illness anorexia research is also relevant to the understanding of diabetes.

Eating, Obesity and Type 2 Diabetes Mellitus

The connection between eating and diabetes follows from the fact that overeating leads to overweight and obesity, which is the major risk factor for the development of insulin resistance and type 2 diabetes mellitus (T2DM). The relationship between eating and diabetes, however, is more complex than this simple unidirectional pathophysiological sequence. Instead, there appear to be positive-feedback links between increased eating, increased adiposity, and insulin resistance, thus setting up vicious cycles that exacerbate diabetes risk or diabetes per se. For example, as shown in figure 12, several vicious cycles apparently result from increased fat intake. (1) Animals with experimentally induced T1DM select and eat fat-containing food in order to obtain utilizable energy [264]. (2) Fat intake may result in brain insulin resistance, i.e. a reduced influence of insulin on eating, which would further increase eating [76]. (3) Increased fat intake per se may stimulate an immune response, which can lead to insulin resistance and thus set up another vicious cycle. Fat metabolism is thought to stimulate the immune system because saturated long-chain fatty acids are structurally similar to bacterial pathogens such as LPS and may therefore target innate immune receptors [265].

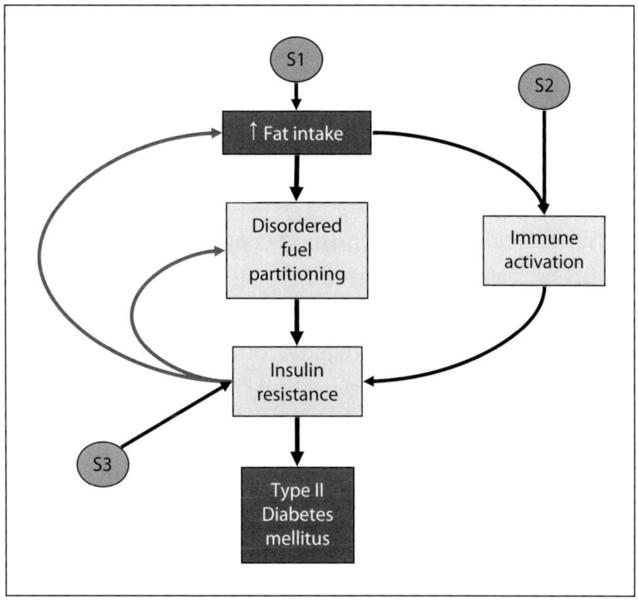

Fig. 12. Schematic of the pathological consequences of increased fat ingestion, emphasizing the positive-feedback or vicious cycle relationships. S1, S2 and S3 indicate potential starting points for these positive-feedback cycles. See text for further details.

Although we have emphasized the relationships between increased ingestion of fat and obesity, it is important to note that causes of increased adiposity other than increased fat intake can set up similar vicious cycles. Finally, recent research increasingly indicates that there are important developmental aspects to these processes. Early development in particular appears to alter metabolic and neural mechanisms in ways that may last a lifetime. The chapters by Bouret [144] and Sullivan and Grove [145] describe some of this work.

Genetics of Eating and Body Weight Phenotypes

Body weight is a highly heritable, polygenetic trait, similar to height. Depending on the measurement method used, the heritability (h^2, the percent of variation in a population phenotype that is due to genetic variation) of body weight or BMI is generally between 0.65 and 0.85 [266]. For example, h^2 of BMI estimated from a comparison of monozygotic and dizygotic Danish male twins was 0.77–0.84, depending on the age at which BMIs were compared [267]. Similarly, adiposity in a group of adult Danes who had been adopted at an early age was strongly and significantly related to the adiposity of their biological parents, but was not significantly related to the adiposity of their adoptive parents [268]. Which genes contribute most to the heritability of obesity remains unclear. According to a recent large (>90,000 total subjects) genomewide association study, eight contributing genes have been identified [269].

It is often asked that if adiposity is highly heritable, how can the prevalence of obesity increase so rapidly without a change in our genome? Two points explain this apparent paradox. First, h^2 measures heritability in a single environment. If the environment changes, h^2 may as well. Thus, individual estimates of h^2 in the present high obesity-risk environment and in the former low obesity-risk environment might each be near 0.8, but this figure would be markedly reduced if h^2 were estimated in both environments simultaneously. Second, although the genome has not changed much in recent times, the particular genes that are expressed and their degree of expression probably have. That is, different genes are likely to effectively contribute to our phenotypes in the present environment more than in former environments. Genes related to the intake of sugar and fat, for example, are certainly more effective regulators of gene expression now than in former days when sugar and fat intakes were lower.

A more directly relevant question in the present context is, are food selection and amount eaten genetically determined traits that contribute to the risk of obesity or other disorders of eating? As recently reviewed by de Krom et al. [270], an increasing number of studies indicate that the answer is yes. Furthermore, allelic variants that contribute to this heritability have been identified. The earliest such reports were rare cases of single-gene mutations that produce dramatic hyperphagia and obesity syndromes in affected people, for example, in individuals lacking leptin due to null mutations in *LEP* [271] or lacking POMC-derived peptides due to null mutations in *pomc* [272].

More interesting are reports of altered eating in relatively more common allelic variants. So far, most is known about the melanocortin-4 receptor gene *MC4R* [266, 269, 273, 274]. The role of MC4R in the control of eating was reviewed above. About 100 *MC4R* variants have been associated with obesity and occur in about 2–6% of obese people, depending on the population sampled. In one study of 17 children with *MC4R* mutations and severe early onset obesity, the degree of impairment in MC4R signaling in vitro was associated with amount eaten during an ad libitum meal, with the energy intake of the most affected children about four times that of controls [275]. A recent analysis of semiquantitative eating questionnaires taken during a longitudinal study of >5,000 US nurses, all American women of European ancestry, revealed that the single-nucleotide polymorphism (SNP) rs17782313 near *MC4R* was significantly associated with increased intakes of energy, fat and protein [276]. This SNP was also associated with BMI, weight gain between ages 44 and 54 years, and T2DM risk in these women. As this SNP was quite common (minor allele frequency of about 25% in this sample), its potential role in human obesity deserves further analysis. Further associations of variants in obesity-related genes and measures of eating behavior and food reward are described in the chapters by Hetherington and Cecil [277], which includes a discussion of the eating effects of variants in the FTO gene, and by Stice and Dagher [4], who focus on genetic variations in dopaminergic food reward.

Finally, an interesting alternative method, the extreme discordant phenotype approach, deserves mention. Beginning with a sample of >17,000 Dutch women, de Krom et al. [278] first identified women who were both obese and in the top 5% of either meal frequency (n = 60) or meal size (n = 72), estimated by questionnaires and the use of photos of food portions, respectively, and then searched for the presence of known SNP in *cck*, *LEP* or *LR*. Women with *LEP* SNP rs4731413 had markedly increased risk for extremely frequent meals, whereas those with CCK SNP rs6801844 had increased risk for extremely large meals. Furthermore, the prevalences of these eating-related SNP were remarkably high in both control and affected women, 58 and 68%, respectively, in the case of the *LEP* SNP. These data indicate not only that specific high-risk eating traits have genetic bases but also that the genes conferring such risks are surprisingly common.

Conclusion

This chapter has presented an overview of the present status of the physiology of eating, in particular as it relates to body weight regulation and the pathophysiology of obesity. As mentioned at the beginning, it also serves as an introduction to the chapters describing specific research frontiers in the physiology of eating.

Dealing with the current obesity epidemic is a major societal problem, and responses at all levels are being sought to reduce the incidence of obesity and to treat obese persons. Overeating and lack of physical activity are recognized as the main causes of this problem. It is increasingly apparent that a multifacetted approach is required to reverse the obesity epidemic, involving numerous adjustments in our culture as well as improved medical approaches. As part of the latter, a better understanding of the physiological controls of eating, in particular as they interrelate to the regulation of body weight and adiposity, would facilitate development of more effective treatments for obesity. As emphasized by De Kloet and Woods [279], at present only two prescription medications are available for the treatment of obesity, but their efficacies are very modest and patients' perception of the quality of life benefit they yield is minimal. Gastric-bypass bariatric surgery is certainly much more effective than any currently available medication, but because of its drawbacks and risks, it is presently considered appropriate for only the persons at the highest risk. From a basic research perspective, the marked decrease in eating after bariatric surgery poses a problem and, perhaps, an opportunity. That is, could the physiological mechanisms that bring about the decreases in eating after bariatric surgery be harnessed in non-surgical ways to develop new treatments? Finally, bariatric surgery also exemplifies the often overlooked point that translational research means not only translating basic research into clinical practice, but also translating clinical experience back into new directions in basic research.

References

1 James W: The Principles of Psychology. London, Dover Press, 1878.

2 Neary NT, Batterham RL: Gaining new insights into food reward with functional neuroimaging; in Langhans W, Geary N (eds): Frontiers in Eating and Weight Regulation. Forum Nutr. Basel, Karger, 2010, vol 63, pp 152–163.

3 Kringelbach ML, Stein A: Cortical mechanisms of human eating; in Langhans W, Geary N (eds): Frontiers in Eating and Weight Regulation. Forum Nutr. Basel, Karger, 2010, vol 63, pp 164–175.

4 Stice E, Dagher A: Genetic variation in dopaminergic reward in humans; in Langhans W, Geary N (eds): Frontiers in Eating and Weight Regulation. Forum Nutr. Basel, Karger, 2010, vol 63, pp 176–185.

5 Hillebrand JJG, Geary N: Do leptin and insulin signal adiposity; in Langhans W, Geary N (eds): Frontiers in Eating and Weight Regulation. Forum Nutr. Basel, Karger, 2010, vol 63, pp 111–122.

6 Kringelbach ML: The human orbitofrontal cortex: Linking reward to hedonic experience. Nature Rev Neurosci 2005;6:691–702.

7 Davis JD, Smith GP: The Conditioned Satiating Effect of Orosensory Stimuli. Physiol Behav 2009.

8 Klein DA, Schebendach JS, Devlin MJ, Smith GP, Walsh BT: Intake, sweetness and liking during modified sham feeding of sucrose solutions. Physiol Behav 2006;87:602–606.

9 Berridge KC, Robinson TE: Parsing reward. Trends Neurosci 2003;26:507–513.

10 Gibson EL, Brunstrom JM: Learned influences on appetite and food intake: evidence in human beings; in Cooper SJ, Kirkham TC (eds): Appetite and Body Weight: Integrative Systems and the Development of Anti-Obesity Drugs. London, Academic Press/Elsevier, 2007, pp 271–300.

11 Rozin PN, Schulkin J: Food selection; in Stricker E (ed): Handbook of Behavioral Neurobiology. New York, Plenum Press, 1990, vol 10, Neurobiology of Food and Fluid Intake, pp 297–328.

12 Bowen DJ: Taste and food preference changes across the course of pregnancy. Appetite 1992;19:233–242.

13 Zellner DA, Garriga-Trillo A, Centeno S, Wadsworth E: Chocolate craving and the menstrual cycle. Appetite 2004;42:119–121.

14 Rozin P, Hanko K, Durlach P: Self-prediction of hedonic trajectories for repeated use of body products and foods: poor performance, not improved by a full generation of experience. Appetite 2006;46: 297–303.

15 Bartoshuk LM, Duffy VB, Hayes JE, Moskowitz HR, Snyder DJ: Psychophysics of sweet and fat perception in obesity: problems, solutions and new perspectives. Phil Trans Royal Soc [B] 2006;361:1137–1148.

16 Drewnowski A, Kurth C, Holden-Wiltse J, Saari J: Food preferences in human obesity: carbohydrates versus fats. Appetite 1992;18:207–221.

17 Macdiarmid JI, Vail A, Cade JE, Blundell JE: The sugar-fat relationship revisited: differences in consumption between men and women of varying BMI. Int J Obes Relat Metab Disord 1998;22:1053–1061.

18 Rolls BJ: Sensory-specific satiety. Nutr Rev 1986;44: 93–101.

19 Rolls BJ, Vanduijvenvoorde PM, Rowe EA: Variety in the diet enhances intake in a meal and contributes to the development of obesity in the rat. Physiol Behav 1983;31:21–27.

20 Banks WA: Blood-brain barrier as a regulatory interface; in Langhans W, Geary N (eds): Frontiers in Eating and Weight Regulation. Forum Nutr. Basel, Karger, 2010, vol 63, pp 102–110.

21 Geary N: Endocrine controls of eating: CCK, leptin, and ghrelin. Physiol Behav 2004;81:719–733.

22 Gibbs J, Young RC, Smith GP: Cholecystokinin decreases food intake in rats. J Comp Physiol Psychol 1973;84:488–495.

23 Smith GP: Introduction to the reviews on peptides and the control of food intake and body weight. Neuropeptides 1999;33:323–328.

24 Beglinger C, Degen L: Gastrointestinal satiety signals in humans: physiologic roles for GLP-1 and PYY? Physiol Behav 2006;89:460–464.

25 Cummings DE, Overduin J: Gastrointestinal regulation of food intake. J Clin Invest 2007;117:13–23.

26 Moran TH: Gut peptide signaling in the controls of food intake. Obesity 2006;14:250–253.

27 Murphy KG, Bloom SR: Gut hormones and the regulation of energy homeostasis. Nature 2006;444:854–859.

28 Ritter RC: Gastrointestinal mechanisms of satiation for food. Physiol Behav 2004;81:249–273.

29 Strader AD, Woods SC: Gastrointestinal hormones and food intake. Gastroenterology 2005;128:175–191.

30 Wölnerhanssen B, Beglinger C: Therapeutic potential of gut peptides; in Langhans W, Geary N (eds): Frontiers in Eating and Weight Regulation. Forum Nutr. Basel, Karger, 2010, vol 63, pp 54–63.

31 Kojima M, Hosoda H, Date Y, Nakazato M, Matsuo H, Kangawa K: Ghrelin is a growth-hormone-releasing acylated peptide from stomach. Nature 1999;402:656–660.

32 Cummings DE, Foster-Schubert KE, Overduin J: Ghrelin and energy balance: Focus on current controversies. Current Drug Targets 2005;6:153–169.

33 Ueno H, Yamaguchi H, Kangawa K, Nakazato M: Ghrelin: a gastric peptide that regulates food intake and energy homeostasis. Regul Peptides 2005;126: 11–19.

34 Zhao HY: Growth hormone secretagogue receptor antagonists as potential therapeutic agents for obesity. Drug Dev Res 2005;65:50–54.

35 Helmling S, Maasch C, Eulberg D, Buchner K, Schroder W, Lange C, et al: Inhibition of ghrelin action in vitro and in vivo by an RNA-Spiegelmer. Proc Natl Acad Sci USA 2004;101:13174–13179.

36 Monnikes H, Helmling S, Stengel A, Wlotzka B, Andresen V, Klapp BF, et al: Anti-ghrelin Spiegelmer NOX-B11 inhibits neurostmoulatory and crexigenic effects of peripheral ghrelin in rats. Gastroenterology 2005;128:A374.

37 Wang SP, Helmling S, Stribling DS, et al: Ghrelin neutralization by an RNA-Spiegelmer yields body weight loss in diet-induced obese mice. Obes Res 2005;13:A32.

38 Gardiner JV, Kong WM, Ward H, Murphy KG, Dhillo WS, Bloom SR: AAV mediated expression of anti-sense neuropeptide Y cRNA in the arcuate nucleus of rats results in decreased weight gain and food intake. Biochem Biophys Res Commun 2005; 327:1088–1093.

39 Gualillo O, Lago F, Dieguez C: Introducing GOAT: a target for obesity and anti-diabetic drugs? Trends Pharmacol Sci 2008;29:398–401.

40 Kirchner H, Gutierrez JA, Solenberg PJ, Pfluger PT, Czyzyk TA, Willency JA, et al: GOAT links dietary lipids with the endocrine control of energy balance. Nat Med 2009;15:741–745.

41 Yang J, Brown MS, Liang G, Grishin NV, Goldstein JL: Identification of the acyltransferase that octanoylates ghrelin, an appetite-stimulating peptide hormone. Cell 2008;132:387–396.

42 Date Y, Murakami N, Toshinai K, Matsukura S, Niijima A, Matsuo H, et al: The role of the gastric afferent vagal nerve in ghrelin-induced feeding and growth hormone secretion in rats. Gastroenterology 2002;123:1120–1128.

43 Date Y, Toshinai K, Koda S, Miyazato M, Shimbara T, Tsuruta T, et al: Peripheral interaction of ghrelin with cholecystokinin on feeding regulation. Endocrinology 2005;146:3518–3525.

44 Arnold M, Mura A, Langhans W, Geary N: Gut vagal afferents are not necessary for the eating-stimulatory effect of intraperitoneally injected Ghrelin in the rat. J Neurosci 2006;26:11052–11060.

45 Cowley MA, Grove KL: Ghrelin-satisfying a hunger for the mechanism. Endocrinology 2004;145:2604–2606.

46 Eisen S, Davis JD, Rauhofer E, Smith GP: Gastric negative feedback produced by volume and nutrient during a meal in rats. Am J Physiol 2001;281: R1201–R1214.

47 Kaplan JMM, Moran TH: Gastrointestinal signaling in the control of food intake; in Stricker EM, Woods SC (eds): Handbook of Behavioral Neurobiology. Neurobiology of Food and Fluid Intake, ed 2. New York, Kluwer Academic, 2004, vol 14, pp 275–305.

48 Phillips RJ, Powley TL: Gastric volume rather than nutrient content inhibits food intake. Am J Physiol 1996;271:R766–R779.

49 Schwartz GJ: Brainstem integrative function in the central nervous system control of food intake; in Langhans W, Geary N (eds): Frontiers in Eating and Weight Regulation. Forum Nutr. Basel, Karger, 2010, vol 63, pp 141–151.

50 Geliebter A, Westreich S, Gage D: Gastric distention by balloon and test-meal intake in obese and lean subjects. Am J Clin Nutr 1988;48:592–594.

51 Rigaud D, Trostler N, Rozen R, Vallot T, Apfelbaum M: Gastric distension, hunger and energy intake after balloon implantation in severe obesity. Int J Obes Relat Metab Disord 1995;19:489–495.

52 Jones KL, Doran SM, Hveem K, et al: Relation between postprandial satiation and antral area in normal subjects. Am J Clin Nutr 1997;66:127–132.

53 Sturm K, Parker B, Wishart J, et al: Energy intake and appetite are related to antral area in healthy young and older subjects. Am J Clin Nutr 2004;80: 656–667.

54 Oesch S, Ruegg C, Fischer B, Degen L, Beglinger C: Effect of gastric distension prior to eating on food intake and feelings of satiety in humans. Physiol Behav 2006;87:903–910.

55 Wang GJ, Tomasi D, Backus W, et al: Gastric distention activates satiety circuitry in the human brain. Neuroimage 2008;39:1824–1831.

56 Beglinger C, Degen L: Fat in the intestine as a regulator of appetite: role of CCK. Physiol Behav 2004; 83:617–621.

57 Moran TH, Ameglio PJ, Schwartz GJ, Mchugh PR: Blockade of type-A, not type-B, CCK receptors attenuates satiety actions of exogenous and endogenous CCK. Am J Physiol 1992;262:R46–R50.

58 Moran TH, Kinzig KP: Gastrointestinal satiety signals. II. Cholecystokinin. Am J Physiol 2004;286: G183–G188.

59 Ballinger AB, Clark ML: L-Phenylalanine releases cholecystokinin (CCK) and is associated with reduced food-intake in humans: evidence for a physiological role of CCK in control of eating. Metab Clin Exp 1994;43:735–738.

60 Lieverse RJ, Jansen JBMJ, Vandezwan A, Samson L, Masclee AAM, Lamers CBHW: Effects of a physiological dose of cholecystokinin on food Intake and postprandial satiation in man. Regul Peptides 1993; 43:83–89.
61 Jordan J, Greenway FL, Leiter LA, Li Z, Jacobson P, Murphy K, et al: Stimulation of cholecystokinin-A receptors with GI181771X does not cause weight loss in overweight or obese patients. Clin Pharmacol Ther 2008;83:281–287.
62 West DB, Fey D, Woods SC: Cholecystokinin persistently suppresses meal size but not food intake in free-feeding rats. Am J Physiol 1984;246:R776–R787.
63 Bi S, Moran TH: Actions of CCK in the controls of food intake and body weight: Lessons from the CCK-A receptor deficient OLETF rat. Neuropeptides 2002;36:171–181.
64 Cox JE: Cholecystokinin satiety involves CCKA receptors perfused by the superior pancreaticoduodenal artery. Am J Physiol 1998;274:R1390–R1396.
65 Blevins JE, Baskin DG: Hypothalamic-brainstem circuits controlling eating; in Langhans W, Geary N (eds): Frontiers in Eating and Weight Regulation. Forum Nutr. Basel, Karger, 2010, vol 63, pp 133–140.
66 Batterham RL, Cowley MA, Small CJ, Herzog H, Cohen MA, Dakin CL, et al: Gut hormone PYY3–36 physiologically inhibits food intake. Nature 2002;418:650–654.
67 Karra E, Chandarana K, Batterham RL: The role of peptide YY in appetite regulation and obesity. J Physiol 2009;587:19–25.
68 Neary MT, Batterham RL: Peptide YY: food for thought. Physiol Behav 2009;97:616–619.
69 Vincent RP, le Roux CW: The satiety hormone peptide YY as a regulator of appetite. J Clin Pathol 2008; 61:548–552.
70 Ruttimann EB, Arnold M, Hillebrand JJ, Geary N, Langhans W: Intrameal hepatic portal and intraperitoneal infusions of glucagon-like peptide-1 reduce spontaneous meal size in the rat via different mechanisms. Endocrinology 2009;150:1174–1181.
71 Williams DL, Baskin DG, Schwartz MW: Evidence that intestinal glucagon-like peptide-1 plays a physiological role in satiety. Endocrinology 2009;150: 1680–1687.
72 Hayes MR, Skibicka KP, Grill HJ: Caudal brainstem processing is sufficient for behavioral, sympathetic, and parasympathetic responses driven by peripheral and hindbrain glucagon-like-peptide-1 receptor stimulation. Endocrinology 2008;149:4059–4068.
73 Scott KA, Moran TH: The GLP-1 agonist exendin-4 reduces food intake in nonhuman primates through changes in meal size. Am J Physiol 2007;293:R983–R987.
74 Sandoval DA, Bagnol D, Woods SC, D'Alessio DA, Seeley RJ: Arcuate glucagon-like peptide 1 receptors regulate glucose homeostasis but not food intake. Diabetes 2008;57:2046–2054.
75 Kim W, Egan JM: The role of incretins in glucose homeostasis and diabetes treatment. Pharmacol Rev 2008;60:470–512.
76 Woods SC, Lutz TA, Geary N, Langhans W: Pancreatic signals controlling food intake; insulin, glucagon and amylin. Phil Trans Royal Soc [B] 2006; 361:1219–1235.
77 Lutz TA: Roles of amylin in satiation, adiposity and brain development; in Langhans W, Geary N (eds): Frontiers in Eating and Weight Regulation. Forum Nutr. Basel, Karger, 2010, vol 63, pp 64–74.
78 Lutz TA, Del Prete E, Scharrer E: Reduction of food intake in rats by intraperitoneal injection of low doses of amylin. Physiol Behav 1994;55:891–895.
79 Lutz TA, Geary N, Szabady MM, Del Prete E, Scharrer E: Amylin decreases meal size in rats. Physiol Behav 1995;58:1197–1202.
80 Lutz TA: Pancreatic amylin as a centrally acting satiating hormone. Curr Drug Targets 2005;6:181–189.
81 Arnelo U, Reidelberger R, Adrian TE, Larsson J, Permert J: Sufficiency of postprandial plasma levels of islet amyloid polypeptide for suppression of feeding in rats. Am J Physiol 1998;275:R1537–R1542.
82 Mollet A, Gilg S, Riediger T, Lutz TA: Infusion of the amylin antagonist AC 187 into the area postrema increases food intake in rats. Physiol Behav 2004;81:149–155.
83 Reidelberger RD, Haver AC, Arnelo U, Smith DD, Schaffert CS, Permert J: Amylin receptor blockade stimulates food intake in rats. Am J Physiol 2004; 287:R568–R574.
84 Langhans W: Metabolic and glucostatic control of feeding. Proc Nutr Soc 1996;55:497–515.
85 Langhans W: The enterocyte as an energy flow sensor in the control of eating; in Langhans W, Geary N (eds): Frontiers in Eating and Weight Regulation. Forum Nutr. Basel, Karger, 2010, vol 63, pp 75–83.
86 Moran TH: Hypothalamic nutrient sensing and energy balance; in Langhans W, Geary N (eds): Frontiers in Eating and Weight Regulation. Forum Nutr. Basel, Karger, 2010, vol 63, pp 94–101.
87 LouisSylvestre J, LeMagnen J: A fall in blood-glucose level precedes meal onset in free-feeding rats. Neurosci Biobehav Rev 1980;4:13–15.
88 Melanson KJ, Westerterp-Plantenga MS, Campfield LA, Saris WHM: Blood glucose and meal patterns in time-blinded males, after aspartame, carbohydrate, and fat consumption, in relation to sweetness perception. Br J Nutr 1999;82:437–446.
89 Campfield LA, Smith FJ: Blood glucose dynamics and control of meal initiation: a pattern detection and recognition theory. Physiol Rev 2003;83:25–58.

90 MacLagan N: The role of appetite in the control of body weight. J Physiol (London) 1937;90:385–394.
91 Bady I, Marty N, Dallaporta M, et al: Evidence from Glut2-null mice that glucose is a critical physiological regulator of feeding. Diabetes 2006;55:988–995.
92 Russek M: Participation of hepatic glucoreceptors in control of intake of food. Nature 1963;197:79–81.
93 Tordoff MG, Friedman MI: Hepatic portal glucose infusions decrease food-intake and increase food preference. Am J Physiol 1986;251:R192–R196.
94 Tordoff MG, Friedman MI: Hepatic control of feeding: effect of glucose, fructose, and mannitol infusion. Am J Physiol 1988;254:R969-R976.
95 Tordoff MG, Tluczek JP, Friedman MI: Effect of hepatic portal glucose-concentration on food-intake and metabolism. Am J Physiol 1989;257:R1474–R1480.
96 Langhans W, Grossmann F, Geary N: Intrameal hepatic-portal infusion of glucose reduces spontaneous meal size in rats. Physiol Behav 2001;73:499–507.
97 Niijima A: Afferent impulse discharges from glucoreceptors in liver of guinea pig. Ann NY Acad Sci 1969;157:690–697.
98 Levin BE: Metabolic sensors: viewing glucosensing neurons from a broader perspective. Physiol Behav 2002;76:397–401.
99 Niijima A: Glucose-sensitive afferent nerve-fibers in the liver and their role in food-intake and blood-glucose regulation. J Auton Nerv Syst 1983;9:207–220.
100 Li B, Xi XC, Roane DS, Ryan DH, Martin RJ: Distribution of glucokinase, glucose transporter GLUT2, sulfonylurea receptor-1, glucagon-like peptide-1 receptor and neuropeptide Y messenger RNAs in rat brain by quantitative real time RT-PCR. Mol Brain Res 2003;113:139–142.
101 Navarro M, de Fonseca FR, Alvarez EA, et al: Colocalization of glucagon-like peptide-1 (GLP-1) receptors, glucose transporter GLUT-2, and glucokinase mRNAs in rat hypothalamic cells: evidence for a role of GLP-1 receptor agonists as an inhibitory signal for food and water intake. J Neurochem 1996;67:1982–1991.
102 Marty N, Bady I, Thorens B: Distinct classes of central GLUT2-dependent sensors control counterregulation and feeding. Diabetes 2006;55:S108–S113.
103 Levin BE: Metabolic sensing neurons and the control of energy homeostasis. Physiol Behav 2006;89:486–489.
104 Leonhardt M, Langhans W: Fatty acid oxidation and control of food intake. Physiol Behav 2004;83:645–651.
105 Dobbins RL, Szczepaniak LS, Bentley B, Esser V, Myhill J, McGarry JD: Prolonged inhibition of muscle carnitine palmitoyltransferase-1 promotes intramyocellular lipid accumulation and insulin resistance in rats. Diabetes 2001;50:123–130.
106 Wanders RJA, Vreken P, den Boer MEJ, Wijburg FA, van Gennip AH, Ijlst L: Disorders of mitochondrial fatty acyl-CoA beta-oxidation. J Inherited Metabol Dis 1999;22:442–487.
107 Wood PA: Genetically modified mouse models for disorders of fatty acid metabolism: pursuing the, nutrigenomics of insulin resistance and type 2 diabetes. Nutrition 2004;20:121–126.
108 Scharrer E: Control of food intake by fatty acid oxidation and ketogenesis. Nutrition 1999;15:704–714.
109 Langhans W: Fatty acid oxidation in the energostatic control of eating: a new idea. Appetite 2008; 51:446–451.
110 Brandt K, Arnold M, Geary N, Langhans W, Leonhardt M: Vagal afferents mediate the feeding response to mercaptoacetate but not to the beta (3) adrenergic receptor agonist CL 316,243. Neurosci Lett 2007;411:104–107.
111 Lam TK, Schwartz GJ, Rossetti L: Hypothalamic sensing of fatty acids. Nat Neurosci 2005;8:579–584.
112 Andersson U, Filipsson K, Abbott CR, Woods A, Smith K, Bloom SR, et al: AMP-activated protein kinase plays a role in the control of food intake. J Biol Chem 2004;279:12005–12008.
113 Minokoshi Y, Alquier T, Furukawa N, et al: AMP-kinase regulates food intake by responding to hormonal and nutrient signals in the hypothalamus. Nature 2004;428:569–574.
114 Cota D, Proulx K, Smith KA, et al: Hypothalamic mTOR signaling regulates food intake. Science 2006;312:927–930.
115 Keesey RE, Hirvonen MD: Body weight set-points: determination and adjustment. Journal of Nutrition 1997;127:S1875–S1883.
116 Langhans W, Geary N: Regulation of body weight; in Munsch S, Beglinger C (eds): Obesity and Binge Eating Disorder. Basel, Karger, 2006, pp 21–40.
117 Kennedy GC: The role of depot fat in the hypothalamic control of food intake in the rat. Proc R Soc Lond [B] 1953;140:578–596.
118 Coleman DL, Hummel KP: Effects of parabiosis of normal with genetically diabetic mice. Am J Physiol 1969;217:1298.
119 Coleman DL: Effects of parabiosis of obese with diabetes and normal mice. Diabetologia 1973;9:294–298.
120 Zhang Y, Proenca R, Maffei M, Barone M, Leopold L, Friedman JM: Positional cloning of the mouse obese gene and its human homologue. Nature 1994; 372:425–432.

121 Chua SC, Chung WK, Wupeng XS, et al: Phenotypes of mouse diabetes and rat fatty due to mutations in the OB (leptin) receptor. Science 1996;271:994–996.
122 Tartaglia LA, Dembski M, Weng X, et al: Identification and expression cloning of a leptin receptor, OB-R. Cell 1995;83:1263–1271.
123 Campfield LA, Smith FJ, Guisez Y, Devos R, Burn P: Recombinant mouse ob protein: evidence for a peripheral signal linking adiposity and central neural networks. Science 1995;269:546–549.
124 Halaas JL, Gajiwala KS, Maffei M, et al: Weight-reducing effects of the plasma-protein encoded by the obese gene. Science 1995;269:543–546.
125 Pelleymounter MA, Cullen MJ, Baker MB, et al: Effects of the obese gene-product on body-weight regulation in ob/ob mice. Science 1995;269:540–543.
126 Ahima RS, Lazar MA: Adipokines and the peripheral and neural control of energy balance. Mol Endocrinol 2008;22:1023–1031.
127 Grill HJ, Schwartz MW, Kaplan JM, Foxhall JS, Breininger J, Baskin DG: Evidence that the caudal brainstem is a target for the inhibitory effect of leptin on food intake. Endocrinology 2002;143:239–246.
128 Malik KF, Young WS: Localization of binding sites in the central nervous system for leptin (OB protein) in normal, obese (ob/ob), and diabetic (db/db) C57BL/6J mice. Endocrinology 1996;137:1497–1500.
129 Eckel LA, Langhans W, Kahler A, Campfield LA, Smith FJ, Geary N: Chronic administration of OB protein decreases food intake by selectively reducing meal size in female rats. Am J Physiol 1998; 275:R186–R193.
130 Kahler A, Geary N, Eckel LA, Campfield LA, Smith FJ, Langhans W: Chronic administration of OB protein decreases food intake by selectively reducing meal size in male rats. Am J Physiol 1998;275:R180–R185.
131 Zhang J, Matheny MK, Tumer N, Mitchell MK, Scarpace PJ: Leptin antagonist reveals that the normalization of caloric intake and the thermic effect of food after high-fat feeding are leptin dependent. Am J Physiol 2007;292:R868–R874.
132 Rosenbaum M, Goldsmith R, Bloomfield D, et al: Low-dose leptin reverses skeletal muscle, autonomic, and neuroendocrine adaptations to maintenance of reduced weight. J Clin Invest 2005;115: 3579–3586.
133 Schwartz MW, Woods SC, Seeley RJ, Barsh GS, Baskin DG, Leibel RL: Is the energy homeostasis system inherently biased toward weight gain? Diabetes 2003;52:232–238.
134 Velkoska E, Morris MJ, Burns P, Weisinger RS: Leptin reduces food intake but does not alter weight regain following food deprivation in the rat. Int J Obes 2003;27:48–54.
135 Woods SC, Decke E, Vasselli JR: Metabolic hormones and regulation of body-weight. Psychol Rev 1974;81:26–43.
136 Woods SC, Seeley RJ: Insulin as an adiposity signal. Int J Obes 2001;25:S35–S38.
137 Bruning JC, Gautam D, Burks DJ, et al: Role of brain insulin receptor in control of body weight and reproduction. Science 2000;289:2122–2125.
138 Woods SC, Seeley RJ, Baskin DG, Schwartz MW: Insulin and the blood-brain barrier. Curr Pharmaceut Design 2003;9:795–800.
139 Benoit SC, Air EL, Coolen LM, et al: The catabolic action of insulin in the brain is mediated by melanocortins. J Neurosci 2002;22:9048–9052.
140 Flynn MC, Scott TR, Pritchard TC, Plata-Salaman CR: Mode of action of OB protein (leptin) on feeding. Am J Physiol 1998;275:R174–R179.
141 Plata-Salaman CR, Oomura Y, Shimizu N: Dependence of food intake on acute and chronic ventricular administration of insulin. Physiol Behav 1986; 37:717–734.
142 Emond M, Schwartz GJ, Ladenheim EE, Moran TH: Central leptin modulates behavioral and neural responsivity to CCK. Am J Physiol 1999;276:R1545–R1549.
143 Riedy CA, Chavez M, Figlewicz DP, Woods SC: Central insulin enhances sensitivity to cholecystokinin. Physiol Behav 1995;58:755–760.
144 Bouret SG: Development of hypothalamic neural networks controlling appetite; in Langhans W, Geary N (eds): Frontiers in Eating and Weight Regulation. Forum Nutr. Basel, Karger, 2010, vol 63, pp 84–93.
145 Sullivan EL, Grove KL: Metabolic imprinting in obesity; in Langhans W, Geary N (eds): Frontiers in Eating and Weight Regulation. Forum Nutr. Basel, Karger, 2010, vol 63, pp 186–194.
146 Anand BK, Brobeck JR: Localization of a feeding center in the hypothalamus of the rat. Proc Soc Exp Biol Med 1951;77:323–324.
147 Hetherington AW, Ranson SW: Hypothalamic lesions and adiposity in the rat. Anatom Rec 1940; 78:149–172.
148 Stellar E: The physiology of motivation. Psychol Rev 1954;61:5–22.
149 Duggan JP, Booth DA: Obesity, Overeating, and rapid gastric-emptying in rats with ventromedial hypothalamic-lesions. Science 1986;231:609–611.
150 Epstein AN, Teitelbaum P: Specific loss of hypoglycemic control of feeding in recovered lateral rats. Am J Physiol 1967;213:1159–1167.

151 Powley TL: Ventromedial hypothalamic syndrome, satiety, and a cephalic phase hypothesis. Psychol Rev 1977;84:89–126.
152 Bernardis LL, Bellinger LL: The lateral hypothalamic area revisited: neuroanatomy, body-weight regulation, neuroendocrinology and metabolism. Neurosci Biobehav Rev 1993;17:141–193.
153 Leibowitz SF: Paraventricular nucleus: primary site mediating adrenergic-stimulation of feeding and drinking. Pharmacol Biochem Behav 1978;8:163–175.
154 Olney JW: Brain lesions obesity and other disturbances in mice treated with monosodium glutamate. Science 1969;164:719.
155 Dahlstrom A, Fuxe K: The autonomic nervous system and the histochemical fluorescence method for the microscopical localization of catecholamines and serotonin. Brain Res Bull 1999;50:365–367.
156 Hillarp NA, Fuxe K, Dahlstrom A: Demonstration and mapping of central neurons containing dopamine noradrenaline and 5-hydroxytryptamine and their reactions to psychopharmaca. Pharmacol Rev 1966;18:727–741.
157 Simerly RB: Anatomical substrates of hypothalamic integration; in Paxinos G (ed): The Rat Nervous System. Amsterdam, Elsevier, 2004, pp 335–368.
158 Swanson LW: The neuroanatomy revolution of the 1970s and the hypothalamus. Brain Res Bull 1999; 50:397.
159 Leibowitz SF: Reciprocal hunger-regulating circuits involving alpha-and beta-adrenergic receptors located, respectively, in ventromedial and lateral hypothalamus. Proc Natl Acad Sci USA 1970;67:1063.
160 Ungerstedt U: Adipsia and aphagia after 6-hydroxydopamine induced degeneration of the nigro-striatal dopamine system. Acta Physiol Scand Suppl 1971;367:95–122.
161 Kirchgessner AL, Sclafani A: PVN-hindbrain pathway involved in the hypothalamic hyperphagia-obesity syndrome. Physiol Behav 1988;42:517–528.
162 Grill HJ, Kaplan JM: The neuroanatomical axis for control of energy balance. Front Neuroendocrinol 2002;23:2–40.
163 Norgren R, Grill HJ: Brain-stem control of ingestive behavior; in Pfaff D (ed): The Physiological Mechanisms of Motivation. New York, Springer, 1982, pp 99–132.
164 Banks WA, Kastin AJ, Huang W, Jaspan JB, Maness LM: Leptin enters the brain by a saturable system independent of insulin. Peptides 1996;17:305–311.
165 Schwartz MW, Seeley RJ, Campfield LA, Burn P, Baskin DG: Identification of targets of leptin action in rat hypothalamus. J Clin Invest 1996;98:1101–1106.
166 Mercer JG, Hoggard N, Williams LM, et al: Coexpression of leptin receptor and preproneuropeptide Y mRNA in arcuate nucleus of mouse hypothalamus. J Neuroendocrinol 1996;8:733–735.
167 Seeley RJ, Yagaloff KA, Fisher SL, et al: Melanocortin receptors in leptin effects. Nature 1997;390:349.
168 Berthoud HR: Multiple neural systems controlling food intake and body weight. Neurosci Biobehav Rev 2002;26:393–428.
169 Cowley MA, Cone RD, Enriori P, Louiselle I, Williams SM, Evans AE: Electrophysiological actions of peripheral hormones on melanocortin neurons. Melanocortin Syst 2003;994:175–186.
170 Ellacott KLJ, Cone RD: The central melanocortin system and the integration of short- and long-term regulators of energy homeostasis. Rec Progr Horm Res 2004;59:395–408.
171 Elmquist JK, Coppari R, Balthasar N, Ichinose M, Lowell BB: Identifying hypothalamic pathways controlling food intake, body weight, and glucose homeostasis. J Comp Neurol 2005;493:63–71.
172 Heisler LK, Jobst EE, Sutton GM, et al: Serotonin reciprocally regulates melanocortin neurons to modulate food intake. Neuron 2006;51:239–249.
173 Hillebrand JJG, de Wied D, Adan RAH: Neuropeptides, food intake and body weight regulation: a hypothalamic focus. Peptides 2002;23:2283–2306.
174 Craig AD: How do you feel? Interoception: the sense of the physiological condition of the body. Nature Rev Neurosci 2002;3:655–666.
175 Rolls ET: Brain mechanisms underlying flavour and appetite. Phil Trans Roy Soc [B] 2006;361:1123–1136.
176 Uher R, Treasure J, Heining M, Brammer MJ, Campbell IC: Cerebral processing of food-related stimuli: effects of fasting and gender. Behav Brain Res 2006;169:111–119.
177 Broberger C, Landry M, Wong H, Walsh JN, Hokfelt T: Subtypes Y1 and Y2 of the neuropeptide Y receptor are respectively expressed in pro-opiomelanocortin- and neuropeptide-Y-containing neurons of the rat hypothalamic arcuate nucleus. Neuroendocrinology 1997;66:393–408.
178 Csiffary A, Gorcs TJ, Palkovits M: Neuropeptide-Y innervation of ACTH-immunoreactive neurons in the arcuate nucleus of rats: a correlated light and electron-microscopic double immunolabeling study. Brain Res 1990;506:215–222.
179 Beck B: Neuropeptide Y in normal eating and in genetic and dietary-induced obesity. Phil Trans Roy Soc [B] 2006;361:1159–1185.
180 Billington CJ, Briggs JE, Grace M, Levine AS: Effects of intracerebroventricular injection of neuropeptide-Y on energy-metabolism. Am J Physiol 1991; 260:R321–R327.

181 Stanley BG, Magdalin W, Seirafi A, Thomas WJ, Leibowitz SF: The perifornical area: the major focus of (a) patchy distributed hypothalamic neuropeptide Y-sensitive feeding system(s). Brain Res 1993; 604:304–317.
182 Widdowson PS: Regionally-selective down-regulation of NPY receptor subtypes in the obese Zucker rat. Relationship to the Y5 'feeding' receptor. Brain Res 1997;758:17–25.
183 Balthasar N, Dalgaard LT, Lee CE, Yu J, Funahashi H, Williams T, et al: Divergence of melanocortin pathways in the control of food intake and energy expenditure. Cell 2005;123:493–505.
184 Heymsfield SB, Greenberg AS, Fujioka K, Dixon RM, Kushner R, Hunt T, et al: Recombinant leptin for weight loss in obese and lean adults: a randomized, controlled, dose-escalation trial. J Am Med Assoc 1999;282:1568–1575.
185 Münzberg H: Leptin-signaling pathways and leptin resistance; in Langhans W, Geary N (eds): Frontiers in Eating and Weight Regulation. Forum Nutr. Basel, Karger, 2010, vol 63, pp 123–132.
186 Fan W, Boston BA, Kesterson RA, Hruby VJ, Cone RD: Role of melanocortinergic neurons in feeding and the agouti obesity syndrome. Nature 1997;385: 165–168.
187 Harrold JA, Widdowson PS, Williams G: Altered energy balance causes selective changes in melanocortin-4 (MC4-R), but not melanocortin-3 (MC3-R), receptors in specific hypothalamic regions: further evidence that activation of MC4-R is a physiological inhibitor of feeding. Diabetes 1999;48:267–271.
188 Mountjoy KG, Mortrud MT, Low MJ, Simerly RB, Cone RD: Localization of the melanocortin-4 receptor (Mc4-R) in neuroendocrine and autonomic control-circuits in the brain. Mol Endocrinol 1994; 8:1298–1308.
189 Moran TH: Neural and hormonal controls of food intake and satiety; in Johnson LR (ed): Physiology of the Gastrointestinal Tract. Burlington, Elsevier Academic Press, 2006, pp 877–894.
190 Campbell RE, Smith MS, Allen SE, Grayson BE, ffrench-Mullen JMH, Grove KL: Orexin neurons express a functional pancreatic polypeptide Y4 receptor. J Neurosci 2003;23:1487–1497.
191 Flynn MC, Turrin NP, Plata-Salaman CR, ffrench-Mullen JMH: Feeding response to neuropeptide Y-related compounds in rats treated with Y5 receptor antisense or sense phosphothio-oligodeoxynucleotide. Physiol Behav 1999;66:881–884.
192 Kanatani A, Ishihara A, Asahi S, Tanaka T, Ozaki S, Ihara M: Potent neuropeptide YY1 receptor antagonist, 1229U91: blockade of neuropeptide Y-induced and physiological food intake. Endocrinology 1996; 137:3177–3182.
193 Kanatani A, Mashiko S, Murai N, et al: Role of the Y1 receptor in the regulation of neuropeptide Y-mediated feeding: comparison of wild-type, Y1 receptor-deficient, and Y5 receptor-deficient mice. Endocrinology 2000;141:1011–1016.
194 Naveilhan P, Hassani H, Canals JM, et al: Normal feeding behavior, body weight and leptin response require the neuropeptide YY2 receptor. Nat Med 1999;5:1188–1193.
195 Blevins JE, Eakin TJ, Murphy JA, Schwartz MW, Baskin DG: Oxytocin innervation of caudal brainstem nuclei activated by cholecystokinin. Brain Res 2003;993:30–41.
196 Blevins JE, Schwartz MW, Baskin DG: Evidence that paraventricular nucleus oxytocin neurons link hypothalamic leptin action to caudal brain stem nuclei controlling meal size. Am J Physiol 2004; 287:R87–R96.
197 Costello JF, Brown MR, Gray TS: Bombesin immunoreactive neurons in the hypothalamic paraventricular nucleus innervate the dorsal vagal complex in the rat. Brain Res 1991;542:77–82.
198 Rinaman L: Oxytocinergic inputs to the nucleus of the solitary tract and dorsal motor nucleus of the vagus in neonatal rats. J Comp Neurol 1998;399:101–109.
199 Broberger C, De Lecea L, Sutcliffe JG, Hokfelt T: Hypocretin/orexin- and melanin-concentrating hormone-expressing cells form distinct populations in the rodent lateral hypothalamus: relationship to the neuropeptide Y and Agouti gene-related protein systems. J Comp Neurol 1998;402:460–474.
200 Chen CT, Dun SL, Kwok EH, Dun NJ, Chang JK: Orexin A-like immunoreactivity in the rat brain. Neurosci Lett 1999;260:161–164.
201 Elias CF, Saper CB, Maratos-Flier E, et al: Chemically defined projections linking the mediobasal hypothalamus and the lateral hypothalamic area. J Comp Neurol 1998;402:442–459.
202 Qu DQ, Ludwig DS, Gammeltoft S, et al: A role for melanin-concentrating hormone in the central regulation of feeding behaviour. Nature 1996;380:243–247.
203 Borowsky B, Durkin MM, Ogozalek K, et al: Antidepressant, anxiolytic and anorectic effects of a melanin-concentrating hormone-1 receptor antagonist. Nature Med 2002;8:825.
204 Marsh DJ, Weingarth DT, Novi DE, et al: Melanin-concentrating hormone 1 receptor-deficient mice are lean, hyperactive, and hyperphagic and have altered metabolism. Proc Natl Acad Sci USA 2002; 99:3240–3245.
205 Haynes AC, Jackson B, Overend P, et al: Effects of single and chronic intracerebroventricular administration of the orexins on feeding in the rat. Peptides 1999;20:1099–1105.

206 Sakurai T: Roles of orexins in regulation of feeding and wakefulness. Neuroreport 2002;13:987–995.

207 Obici S, Feng ZH, Morgan Y, Stein D, Karkanias G, Rossetti L: Central administration of oleic acid inhibits glucose production and food intake. Diabetes 2002;51:271–275.

208 Routh VH: Glucose-sensing neurons: are they physiologically relevant? Physiol Behav 2002;76:403–413.

209 Cai XJ, Widdowson PS, Harrold J, et al: Hypothalamic orexin expression: modulation by blood glucose and feeding. Diabetes 1999;48:2132–2137.

210 Williams G, Harrold JA, Cutler DJ: The hypothalamus and the regulation of energy homeostasis: lifting the lid on a black box. Proc Nutr Soc 2000;59: 385–396.

211 van den Pol AN, Gao XB, Obrietan K, Kilduff TS, Belousov AB: Presynaptic and postsynaptic actions and modulation of neuroendocrine neurons by a new hypothalamic peptide, hypocretin/orexin. J Neurosci 1998;18:7962–7971.

212 Moran TH, Bi S: Hyperphagia and obesity of OLETF rats lacking CCK1 receptors: developmental aspects. Dev Psychobiol 2006;48:360–367.

213 Dhillon H, Zigman JM, Ye CP, et al: Leptin directly activates SF1 neurons in the VMH, and this action by leptin is required for normal body-weight homeostasis. Neuron 2006;49:191–203.

214 Xu BJ, Goulding EH, Zang KL, et al: Brain-derived neurotrophic factor regulates energy balance downstream of melanocortin-4 receptor. Nature Neurosci 2003;6:736–742.

215 Dumont Y, Jacques D, Bouchard P, Quirion R: Species differences in the expression and distribution of the neuropeptide Y Y1, Y2, Y4, and Y5 receptors in rodents, guinea pig, and primates brains. J Comp Neurol 1998;402:372–384.

216 Fodor M, Sluiter A, FrankhuijzenSierevogel A, et al: Distribution of Lys-gamma(2)-melanocyte-stimulating hormone-(Lys-gamma(2)-MSH)-like immunoreactivity in neuronal elements in the brain and peripheral tissues of the rat. Brain Res 1996;731:182–189.

217 Shughrue PJ, Lane MV, Merchenthaler I: Comparative distribution of estrogen receptor-alpha and -beta mRNA in the rat central nervous system. J Comp Neurol 1997;388:507–525.

218 Faulconbridge LF, Cummings DE, Kaplan JM, Grill HJ: Hyperphagic effects of brainstem ghrelin administration. Diabetes 2003;52:2260–2265.

219 Grill HJ, Kaplan JM: Sham feeding in intact and chronic decerebrate rats. Am J Physiol 1992;262:R1070–R1074.

220 Grill HJ, Smith GP: Cholecystokinin decreases sucrose intake in chronic decerebrate rats. Am J Physiol 1988;254:R853–R856.

221 Kaplan JM, Seeley RJ, Grill HJ: Daily caloric-intake in intact and chronic decerebrate rats. Behav Neurosci 1993;107:876–881.

222 Scott TR, Giza BK: Issues of gustatory neural coding: Where they stand today. Physiol Behav 2000;69: 65–76.

223 Asarian L, Geary N: Estradiol enhances cholecystokinin-dependent lipid-induced satiation and activates estrogen receptor-alpha-expressing cells in the nucleus tractus solitarius of ovariectomized rats. Endocrinology 2007;148:5656–5666.

224 Thammacharoen S, Lutz TA, Geary N, Asarian L: Hindbrain administration of estradiol inhibits feeding and activates estrogen receptor-alpha-expressing cells in the nucleus tractus solitarius of ovariectomized rats. Endocrinology 2008;149:1609–1617.

225 Barrachina MD, Martinez V, Wang LX, Wei JY, Tache Y: Synergistic interaction between leptin and cholecystokinin to reduce short-term food intake in lean mice. Proc Natl Acad Sci USA 1997;94:10455–10460.

226 Berridge KC, Kringelbach ML: Affective neuroscience of pleasure: reward in humans and animals. Psychopharmacol (Berl) 2008;199:457–480.

227 Berridge KC: Motivation concepts in behavioral neuroscience. Physiol Behav 2004;81:179–209.

228 Nestler EJ, Carlezon WA: The mesolimbic dopamine reward circuit in depression. Biol Psychiatry 2006;59:1151–1159.

229 Zheng HY, Corkern M, Stoyanova I, Patterson LM, Tian R, Berthoud HR: Peptides that regulate food intake: appetite-inducing accumbens manipulation activates hypothalamic orexin neurons and inhibits POMC neurons. Am J Physiol 2003;284:R1436–R1444.

230 Kelley AE, Baldo BA, Pratt WE, Will MJ: Corticostriatal-hypothalamic circuitry and food motivation: Integration of energy, action and reward. Physiol Behav 2005;86:773–795.

231 Norgren R, Hajnal A, Mungarndee SS: Gustatory reward and the nucleus accumbens. Physiol Behav 2006;89:531–535.

232 Schultz W: Behavioral theories and the neurophysiology of reward. Ann Rev Psychol 2006;57:87–115.

233 Wise RA: Role of brain dopamine in food reward and reinforcement. Phil Trans Roy Soc B 2006;361: 1149–1158.

234 Hajnal A, Smith GP, Norgren R: Oral sucrose stimulation increases accumbens dopamine in the rat. Am J Physiol 2004;286:R31–R37.

235 Liang NC, Hajnal A, Norgren R: Sham feeding corn oil increases accumbens dopamine in the rat. Am J Physiol 2006;291:R1236–R1239.

236 Zhang M, Kelley AE: Enhanced intake of high-fat food following striatal mu-opioid stimulation: microinjection mapping and Fos expression. Neuroscience 2000;99:267–277.

237 Zhang M, Balmadrid C, Kelley AE: Nucleus accumbens opioid, GABAergic, and dopaminergic modulation of palatable food motivation: contrasting effects revealed by a progressive ratio study in the rat. Behav Neurosci 2003;117:202–211.
238 Yeomans MR, Wright P, Macleod HA, Critchley JAJH: Effects of nalmefene on feeding in humans: dissociation of hunger and palatability. Psychopharmacology 1990;100:426–432.
239 Kirkham TC: Endocannabinoids in the regulation of appetite and body weight. Behav Pharmacol 2005; 16:297–313.
240 Balleine BW, Killcross AS, Dickinson A: The effect of lesions of the basolateral amygdala on instrumental conditioning. J Neurosci 2003;23:666–675.
241 Kim EM, Quinn JG, Levine AS, O'Hare E: A bidirectional mu-opioid-opioid connection between the nucleus of the accumbens shell and the central nucleus of the amygdala in the rat. Brain Res 2004; 1029:135–139.
242 Will MJ, Franzblau EB, Kelley AE: The amygdala is critical for opioid-mediated binge eating of fat. Neuroreport 2004;15:1857–1860.
243 Stratford TR, Kelley AE, Simansky KJ: Blockade of GABA(A) receptors in the medial ventral pallidum elicits feeding in satiated rats. Brain Res 1999;825: 199–203.
244 Kelley AE, Baldo BA, Pratt WE: A proposed hypothalamic-thalamic-striatal axis for the integration of energy balance, arousal, and food reward. J Comp Neurol 2005;493:72–85.
245 Wizemann TM, Pardue M-L: Exploring the Biological Contribution to Human Health: Does Sex Matter? Washington, National Academy Press, 2001.
246 Asarian L, Geary N: Cyclic estradiol treatment normalizes body weight and restores physiological patterns of spontaneous feeding and sexual receptivity in ovariectomized rats. Horm Behav 2002;42:461–471.
247 Asarian L, Geary N: Modulation of appetite by gonadal steroid hormones. Phil Trans Roy Soc [B] 2006;361:1251–1263.
248 Geary N, Lovejoy J: Sex differences in energy metabolism, obesity and eating behavior; in Becker J, Berkley KJ, Geary N, Hampson E, Herman JP, Young EA (eds): Sex on the Brain: From Genes to Behavior. New York, Oxford University Press, 2007, pp 253–274.
249 Clegg DJ, Brown LM, Kemp CJ, et al: Estradiol-dependent decrease in the orexigenic potency of ghrelin in female rats. Diabetes 2007;56:1051–1058.
250 Eckel LA, Rivera HM, Atchley DP: The anorectic effect of fenfluramine is influenced by sex and stage of the estrous cycle in rats. Am J Physiol 2005;288: R1486–R1491.
251 Musatov S, Chen W, Pfaff DW, et al: Silencing of estrogen receptor alpha in the ventromedial nucleus of hypothalamus leads to metabolic syndrome. Proc Natl Acad Sci USA 2007;104:2501–2506.
252 Shi H, Seeley RJ, Clegg DJ: Sexual differences in the control of energy homeostasis. Front Neuroendocrinol 2009;30:396–404.
253 Wajchenberg BL: Subcutaneous and visceral adipose tissue: their relation to the metabolic syndrome. Endocr Rev 2000;21:697–738.
254 Roepke TA, Xue C, Bosch MA, Scanlan TS, Kelly MJ, Ronnekleiv OK: Genes associated with membrane-initiated signaling of estrogen and energy homeostasis. Endocrinology 2008;149:6113–6124.
255 Tobias JH, Steer CD, Vilarino-Guell C, Brown MA: Effect of an estrogen receptor-alpha intron 4 polymorphism on fat mass in 11-year-old children. J Clin Endocrinol Metab 2007;92:2286–2291.
256 Blaak E: Gender differences in fat metabolism. Curr Opin Clin Nutr Metab Care 2001;4:499–502.
257 Bruns CM, Kemnitz JW: Sex hormones, insulin sensitivity, and diabetes mellitus. Ilar Journal 2004;45: 160–169.
258 Hart BL: Biological basis of the behavior of sick animals. Neurosci Biobehav Rev 1988;12:123–137.
259 Murray MJ, Murray AB: Anorexia of infection as a mechanism of host defense. Am J Clin Nutr 1979; 32:593–596.
260 Rietschel ET, Schletter J, Weidemann B, et al: Lipopolysaccharide and peptidoglycan: CD14-dependent bacterial inducers of inflammation. Microb Drug Resist 1998;4:37–44.
261 Langhans W: Signals generating anorexia during acute illness. Proc Nutr Soc 2007;66:321–330.
262 Miller AH: Norman Cousins Lecture: Mechanisms of cytokine-induced behavioral changes: psychoneuroimmunology at the translational interface. Brain Behav Immun 2009;23:149–158.
263 Park SM, Gaykema RP, Goehler LE: How does immune challenge inhibit ingestion of palatable food? Evidence that systemic lipopolysaccharide treatment modulates key nodal points of feeding neurocircuitry. Brain Behav Immun 2008.
264 Friedman MI, Ramirez I, Edens NK, Granneman J: Food-intake in diabetic rats: isolation of primary metabolic effects of fat feeding. Am J Physiol 1985; 249:R44–R51.
265 Lee JY, Sohn KH, Rhee SH, Hwang D: Saturated fatty acids, but not unsaturated fatty acids, induce the expression of cyclooxygenase-2 mediated through Toll-like receptor 4. J Biol Chem 2001;276:16683–16689.
266 O'Rahilly S, Farooqi IS: Genetics of obesity. Phil Trans Roy Soc [B] 2006;361:1095–1105.
267 Stunkard AJ, Foch TT, Hrubec Z: A twin study of human obesity. J Am Med Ass 1986;256:51–54.

268 Stunkard AJ, Sorensen TI, Hanis C, et al: An adoption study of human obesity. N Engl J Med 1986; 314:193–198.
269 Willer CJ, Speliotes EK, Loos RJ, et al: Six new loci associated with body mass index highlight a neuronal influence on body weight regulation. Nat Genet 2009; 41:25–34.
270 de Krom M, Bauer F, Collier D, Adan RA, La Fleur SE: Genetic variation and effects on human eating behavior. Annu Rev Nutr 2009;29:283–304.
271 Montague CT, Farooqi IS, Whitehead JP, et al: Congenital leptin deficiency is associated with severe early-onset obesity in humans. Nature 1997; 387:903–908.
272 Krude H, Biebermann H, Luck W, Horn R, Brabant G, Gruters A: Severe early-onset obesity, adrenal insufficiency and red hair pigmentation caused by POMC mutations in humans. Nat Genet 1998;19: 155–157.
273 Adan RA, Tiesjema B, Hillebrand JJ, La Fleur SE, Kas MJ, de Krom M: The MC4 receptor and control of appetite. Br J Pharmacol 2006;149:815–827.
274 Kublaoui BM, Zinn AR: Editorial: MC4R mutations – weight before screening! J Clin Endocrinol Metab 2006;91:1671–1672.
275 Farooqi IS, Keogh JM, Yeo GSH, Lank EJ, Cheetham T, O'Rahilly S: Clinical spectrum of obesity and mutations in the melanocortin 4 receptor gene. N Engl J Med 2003;348:1085–1095.
276 Qi L, Kraft P, Hunter DJ, Hu FB: The common obesity variant near MC4R gene is associated with higher intakes of total energy and dietary fat, weight change and diabetes risk in women. Hum Mol Genet 2008;17:3502–3508.
277 Hetherington MM, Cecil JE: Gene-environment interactions in obesity; in Langhans W, Geary N (eds): Frontiers in Eating and Weight Regulation. Forum Nutr. Basel, Karger, 2010, vol 63, pp 195–203.
278 de Krom M, van der Schouw YT, Hendriks J, et al: Common genetic variations in CCK, leptin, and leptin receptor genes are associated with specific human eating patterns. Diabetes 2007;56:276–280.
279 De Kloet AD, Woods SC: Obesity and food intake: basic and clinical approaches; in Langhans W, Geary N (eds): Frontiers in Eating and Weight Regulation. Forum Nutr. Basel, Karger, 2010, vol 63, pp 1–8.
280 Langhans W, Harrold J, Williams G, Geary N: Control of Eating; in Williams G, Frühbeck G (eds): Obesity – Science to Practice. Oxford, Wiley-Blackwell, 2009, pp 125–163.
281 Langhans W, Geary N, Lutz TA: Steuerung der Nahrungsaufnahme; in von Engelhardt W (ed): Physiologie der Haustiere, Stuttgart, Medizinverlage Stuttgart, 2009, pp 616–630.
282 Liebling DS, Eisner JD, Gibbs J, Smith GP: Intestinal satiety in rats. J Comp Physiol Psychol 1975;89:955–965.

Prof. W. Langhans
Physiology and Behaviour Group, Institute of Food Science, Nutrition and Health,
ETH Zürich, Schorenstrasse 16
Schwerzenbach CH–8603 (Switzerland)
Tel. +41 44 6557420, Fax +41 44 6557206, E-Mail wolfgang-langhans@ethz.ch

Langhans W, Geary N (eds): Frontiers in Eating and Weight Regulation.
Forum Nutr. Basel, Karger, 2010, vol 63, pp 54–63

Therapeutic Potential of Gut Peptides

Bettina Wölnerhanssen[a] · Christoph Beglinger[b]

Departments of [a]Visceral Surgery and [b]Gastroenterology, University Hospital, Basel, Switzerland

Abstract

A great deal of research interest is directed toward understanding the control of appetite and regulation of metabolism. It seems as if an epidemic of obesity is sweeping the world, and type II diabetes (T2DM) is following in its wake. The regulation of energy homeostasis is an area that straddles neurobiology, classical endocrinology and metabolism. It is currently one of the most exciting and rapidly advancing topics in medical research, and is also one of the most frustrating areas. The availability of highly palatable, calorie-dense food, together with the low requirement for physical activity in our modern environment, are major factors contributing to the obesity epidemic. If energy intake exceeds energy use, the excess calories are stored as body fat. Knowledge of the homeostatic system that controls body weight has increased dramatically over the last years and has revealed new potential targets for the treatment of obesity. One therapeutic approach is the development of agents based on the gastrointestinal hormones that control food intake and appetite. This review discusses several gut hormones and ligands for their receptors as potential anti-obesity treatments.

Obesity, defined by a body mass index (BMI; weight (kg) divided by height (m) squared) greater than 30, is a chronic disease associated with multiple morbidities and increased mortality. The International Obesity Taskforce estimates that nearly 1.7 billion people in the world may be at risk for weight-related disease, and current mortality rates (more than 2.5 million deaths per year) due to high BMI are expected to double by the year 2030 [1]. The global impact of obesity can also be described in economic terms: 2–7% of total health care costs in the US and UK are attributed to weight-related disease. The global epidemic of obesity is accompanied by enormous human costs. Strategies to combat overweight and obesity have been deceptively simplified into a single logical imperative – energy expenditure should exceed energy storage. This thermodynamic principle guided efforts to develop anti-obesity pharmacotherapy prior to current clinical appreciation for physiological regulators of appetite and energy balance. Unfortunately, once adipose tissue accumulates, a complex regulatory system involving different neuroendocrine responses opposes its diminution. When food intake is limited, counter-regulatory mechanisms induce an increase in

Table 1. Gut hormones used as drugs

Peptide	Used for the treatment of...	Drug used in clinical trials	Drug approved by FDA	Route of administration
Amylin	diabetes, (obesity)	yes; pramlintide	yes	subcutaneous
Ghrelin	anorexia, cachexia	yes	no	intravenous or oral
GLP-1	diabetes, (obesity)	yes:		
		exenatide,	yes	
		liraglutide,	no (drug approved by European Agency EMEA)	subcutaneous
		albugon, ZP 10	no	
PYY	obesity	yes	no	intranasal, i.v., oral

appetite and a decrease in energy expenditure as protective measures against starvation [2]. Weight loss by hypocaloric diets is therefore difficult to achieve, and other measures including pharmacological intervention are often necessary adjuncts to aid the induction of weight loss and maintenance of a target weight. This review summarizes the potential of specific gastrointestinal peptides involved in the control of eating that may be targeted for new and effective treatments.

Gut Hormones as Satiety Peptides

The gastrointestinal tract is the body's largest endocrine organ, producing a variety of hormones that play important roles in energy homeostasis [3]. Gut hormones act as appetite controllers either through direct actions in the brain or indirectly, via actions in the periphery (inhibition of gastric emptying, metabolic effects, stimulation of peripheral nerves and so on). Most gut hormones are anorexigenic (satiating); only one orexigenic gut hormone – ghrelin – is known so far. In this review, we concentrate on the clinical application and therapeutic potential of four of the most promising gut hormones, amylin, ghrelin, GLP-1 and PYY (table 1).

Amylin

Amylin is co-secreted with insulin, in proportion to nutrient intake, from the beta cells of the pancreas. As reviewed by Lutz [4] and others [5–10], amylin may act both as a satiation signal and as an adiposity signal. It also retards gastric emptying [4]. The satiating action of amylin in rats is apparently mediated by a direct action on the area postrema in the hindbrain [10]. Interestingly, experiments employing amylin antagonists indicate that amylin is required for CCK's satiation effect [10]. In view of amylin's central site of action, this is likely to be an interaction in the central processing of the two peptides' signals. With respect to factors that stimulate amylin secretion and its action as an adiposity signal, the reader is referred to the chapter on 'Amylin' by Thomas Lutz [this vol.].

The peptide amylin forms the basis of pramlintide (Symlin; Amylin Pharmaceuticals), an amylin analog that is used as a novel treatment for T2DM and that has been approved by the US Food and Drug Administration (FDA). Pramlintide has a favorable effect on blood glucose and reduces food intake [11]; a moderate reduction in body weight in overweight subjects has been reported (3.5 kg in 16 weeks). Further evaluation of pramlintide as a specific treatment against obesity is still awaited.

Ghrelin

Ghrelin is a 28-amino acid peptide synthesized mainly by G cells in the fundus of the stomach [12], and it is the only known orexigenic gut hormone [13, 14]. Until recently, clinical research has focused on its use in cachexia and anorexia [15–18]; an antagonist has been used only in preclinical trials in Zucker rats so far [19]. Recent new experimental evidence suggests that the biology of ghrelin might be changing rapidly. An octanyl side chain on its third amino acid, a serine, modifies the biological effects of ghrelin, which is essential for its activity. Yang et al. [20] recently reported the identification of the enzyme catalyzing the addition of this octanyl, ghrelin O-acyltransferase (GOAT). GOAT belongs to a family of membrane-bound O-acyltransferases (MBOATs) [21]. The discovery of GOAT raises important questions and suggests several novel therapeutic targets. Indeed, drugs that inhibit GOAT might be able to prevent diet-induced obesity and might be effective therapies for T2DM because they increase insulin secretion and peripheral insulin sensitivity [22]. Ghrelin exerts its central orexigenic action mainly in the hypothalamus and in particular in the hypothalamic arcuate nucleus (Arc), where it activates specific G-protein-coupled receptors (GHS-R) [23], which leads to the synthesis of neuropeptide Y (NPY) and agouti-related protein (AgRP) and increases food intake [24]. The gastric peptide ghrelin may thus function as part of the orexigenic pathway and provide a potential therapeutic target for obesity.

Several ghrelin analogs that could be used for oral application are in development [25, 26]. Their long-lasting effect on hGH (human growth hormone) release has potential clinical applications. Synthetic growth hormone secretagogues have been shown to exert protective effects on ischemic cardiac muscle, with beneficial effects on myocardial structure and function in patients with chronic heart failure with and without impairment [27, 28]. Initial results also indicate significant improvement of cardiovascular function in those patients after a 3-week treatment with intravenous administration of synthetic human ghrelin [17]. Intravenous ghrelin has also been reported to stimulate gastric emptying of solids and to increase appetite in healthy humans [29]. An intravenous or subcutaneous application is, however, cumbersome and impractical for chronic treatment regimens; clearly preferable would be oral agonists that could overcome the problems associated with the intravenous administration. Whether such agonists would be clinically beneficial remains to be determined.

Glucagon-Like Peptide 1

The circulating hormone glucagon-like peptide 1 (GLP-1) – a 30-amino acid peptide – is a product of proglucagon cleavage. GLP-1 is released by L cells of the distal small intestine postprandially in proportion to the calories ingested [30]. GLP-1 acts mainly as an incretin hormone, promoting postprandial insulin release and improving pancreatic beta cell function [30]. In rodents, injection of GLP-1 either directly into the CNS [31] or peripherally [32–35] results in an inhibition of eating. In obese humans, circulating GLP-1 levels are reduced and postprandial responses to GLP-1 are attenuated [36] compared to what occurs in normal-weight controls. To date, clinical development has concentrated on GLP-1 as an antidiabetic agent. For example, subcutaneous administration of native GLP-1 improved blood glucose levels in poorly controlled T2DM patients [37]. In addition, however, a meta-analysis indicates that GLP-1 also has a significant weight-reducing effect in humans [38].

A major disadvantage of native GLP-1 is its short half-life: Within the circulation, GLP-1 is rapidly inactivated by DPP-IV (dipeptidyl peptidase-4) having a half-life in the order of magnitude of seconds. Due to its short half-life, native GLP-1 is consequently not feasible for long-term clinical use. Because of this, novel classes of pharmaceutical agents have been developed that are suitable for ambulatory treatment. At the moment there are two types of incretin-based drugs that improve the effects of GLP-1 in controlling blood glucose levels: GLP-1 agonists (incretin mimetics) and DPP-IV inhibitors [39]. Incretin mimetics are molecules that mimic the action of native GLP-1 at its receptor (e.g. exenatide, liraglutide), and that overcome some of the shortcomings of the natural compound, especially with regard to biological half-life. Exenatide or liraglutide have half-lives of several hours, and both improve diabetes control and produce weight loss in patients [39]. A recent, randomized, open-label study with 464 T2DM patients compared the efficacy and safety of liraglutide to exenatide [40]. Treatment with liraglutide led to a statistically significantly greater drop in glycated hemoglobin, also called HbA1c (drop in HbA1c of 1.12% compared to 0.79% with exenatide). Glycated hemoglobin is a substance in red blood cells that is formed when blood sugar (glucose) attaches to hemoglobin. A drop in glycated hemoglobin indicates a better metabolic control with a corresponding risk reduction for complications. Liraglutide was also significantly better than exenatide at lowering fasting plasma glucose in this T2DM population [40]. Each treatment led to a weight reduction of about 3 kg during the 26-week study. Nausea was the most commonly reported adverse event with both treatments, but there was less persistent nausea and fewer minor hypoglycemic events with liraglutide compared to exenatide [40]. Exenatide is licensed for use as an adjunctive therapy for suboptimal glucose control in T2DM patients, whereas liraglutide is so far available for treatment in Europe only. In studies with liraglutide and, especially, exenatide it has been observed incidentally that many patients lost weight. In-depth investigations of the effects of these compounds on weight loss, however, have not yet been reported. Dose-limiting side effects of GLP-1 and GLP-1 agonists could be nausea and vomiting [41].

DPP-IV inhibitors are another class of GLP-1-based therapies [42]. Several DPP-IV inhibitors now in development are able to approximately double postprandial circulating levels of total and intact GLP-1. Regulatory authorities in different countries have approved two of them, vildagliptin and sitagliptin, for the treatment of poorly controlled T2DM. Orally administered DPP-IV inhibitors, such as sitagliptin and vildagliptin, reduce HbA1c by 0.5–1.0%, with few adverse effects and no weight gain [42]. They differ from incretin mimetics because they have greater oral bioavailability, elicit fewer side effects with overdose, have no direct CNS effects (nausea and vomiting), and have no effect on body weight. Although they are not yet considered first line therapies, incretin mimetics and enhancers expand the options for T2DM treatment and have even been proposed to be used prophylactically.

Peptide YY

Peptide YY (PYY) is a 36-amino acid peptide co-secreted from intestinal L cells with GLP-1 and oxyntomodulin. Like GLP-1, PYY is released from the gut into the circulation in a nutrient-dependent manner: PYY levels are low in the fasting state, rapidly increase in response to food intake, reach a peak 1–2 h postprandially and then remain elevated for several hours [43]. The N-terminal truncation of PYY by DPP-IV results in the major active circulating form, PYY(1–36) [44, 45]. Although the full-length form binds to all Y receptor subtypes, PYY(3–36) binds to the Y2 and Y5 more selectively [46]. The appetite-inhibitory effects of PYY(3–36) are mediated by presynaptic Y2 receptors in the arcuate nucleus, as the anorexigenic effects of PYY are lost in Y2-null mice [47]. There is debate surrounding the principal site of action of PYY(3–36) in the CNS. Batterham et al. [47] propose the hypothalamus as the main target of circulating PYY3–36, whereas Abbott et al. [48], who observed that vagotomy attenuates the ability of systemically administered PYY(3–36) to induce c-Fos expression in the hypothalamus, suggest that the vagus nerve and its brainstem projection sites may predominate as the site of action of PYY(3–36).

Some studies have demonstrated lower fasting and meal-stimulated circulating levels of PYY in obese individuals compared to their lean counterparts, suggesting that obesity might be a PYY-deficiency state. In clinical trials, intravenous administration of PYY(3–36) resulted in reduced food intake in healthy subjects [47] and in reduced food intake as well as a reduction of body weight in obese subjects [47, 49–53]. Other potentially clinically relevant effects of PYY include increases in energy expenditure and fat oxidation, reported in a human study [55], and improved insulin sensitivity at doses that did not affect body weight in rodent studies [54, 55]. Unfortunately, nausea and vomiting appear to be dose-related side effects of PYY, which could limit its therapeutic potential.

Notwithstanding the narrow therapeutic window between appetite reduction and nausea, PYY(3–36) remains an attractive target for obesity treatment that is being pursued by several pharmaceutical companies. Nastech Pharmaceutical company, in collaboration with Merck, has completed phase 1 trials of PYY(3–36) delivered via an intranasal route. Preprandial nasal administration of PYY(3–36) led to a significant reduction in visual-analog appetite scores and a trend toward a dose-dependent reduction in food intake at a test meal. Furthermore, thrice-daily administration of PYY nasal spray in obese volunteers led to reduction in daily caloric intake and weight loss of 0.6 kg after 6 days of treatment [56]. In another study, PYY(3–36) was administered subcutaneously to obese male subjects. The peptide was able to dose-dependently induce lower hunger and thirst ratings and higher satiation feelings. In view of the reproducibility of prolonged plasma PYY levels, subcutaneous administration of PYY(3–36) also seems to be a suitable route of administration for long-term therapeutic use. Finally, we have recently shown that the oral application of PYY(3–36) is feasible using a new proprietary technique, a delivery agent-based approach [57]. The potential clinical applications include reduction in food intake and promotion of weight loss [56].

Conclusion

At present, no effective nonsurgical treatment of human obesity has been found. However even the slightest effect on body weight can influence co-morbidities associated with obesity – especially T2DM – and in that way increase life quality and life expectancy. Modest weight loss results from the use of the two available weight-loss compounds currently approved by the US FDA, orlistat and sibutramine, and similarly modest results were produced by rimonabant during the brief period in which it was approved [58, 59]. Thus, more effective non-surgical treatment is urgently needed. The mechanisms of satiation and postprandial satiety are still not fully understood. Strategies aimed at modulating gut hormone levels or targeting their receptors as potential therapy in obesity seem very attractive: yet the development of gut peptides as drugs is confronted with a number of problems:

- The *short half-life* of many gut hormones and the poor oral bioavailability of peptides and proteins in general suggests the necessity of parenteral formulations, which (a) do not mimic physiological release, and (b) are inconvenient for the patient and will inevitably lead to compliance problems. The intranasal administration of PYY comprises an interesting way of how to circumvent this problem, yet also is nonphysiological. Oral administration of GLP-1 and PYY using a special technology has also recently been shown to be safe and effective in short-term administration, with the potential for more physiological dosing patterns [56].
- Typical dose-related *adverse effects* of all gut hormones seem to be nausea and vomiting. Unfortunately, satiation and nausea are probably related at the neuroendocrine level, so that the therapeutic window for administration of GI peptides may be narrow. Nausea might be reduced by stepwise escalation of doses [41, 59].
- The *use of single gut hormones* seems nonphysiological. In vivo a number of GI hormones are concomitantly released during and after meals and interact in the control of eating. Combining various GI peptides might be the solution. Combined administration of GLP-1 and PYY inhibited eating additively in humans [14], and combinations of PYY with amylin had favorable interactive effects in rodents [60].
- *Durability* of all reported effects is unknown. Potential long-term benefits need to be studied further.

Acknowledgements

This research is supported by a grant of the Swiss National Science Foundation (grant No. 320000–118330).

References

1 http://www.iotf.org/media/iotfaug25.htm; http://www.iotf.org/database/index.asp
2 Leibel RL: The role of leptin in the control of body weight. Nutr Rev 2002;60:S15–S19.
3 Cummings DE, Overduin J: Gastrointestinal regulation of food intake. Physiol Behav 2006;89:465–471.
4 Lutz TA: Amylinergic control of food intake. J Clin Invest 2007;117:13–23.
5 Young AA, Gedulin, B, Vine W, Percy A, Rink TJ: Gastric emptying is accelerated in diabetic BB rats and is slowed by subcutaneous injections of amylin. Diabetologia 1995;38:642–648.
6 Gedulin BR, Rink TJ, Young AA: Dose-response for glucagonostatic effect of amylin in rats. Metab Clin Exp 1997;46:67–70.
7 Butler PC, Chou J, Carter WB, Wang YN, Bu BH, Chang D, Chang JK, Rizza RA: Effects of meal ingestion on plasma amylin concentration in NIDDM and nondiabetic humans. Diabetes 1990;39:752–756.
8 Lutz TA, Del Prete E, Scharrer E: Reduction of food intake in rats by intraperitoneal injection of low doses of amylin. Physiol Behav 1994;55:891–895.
9 Rushing PA, Hagan MM, Seeley RJ, Lutz TA, Woods SC: Amylin: a novel action in the brain to reduce body weight. Endocrinology 2000;141:850–853.
10 Woods SC, Lutz TA, Geary N, Langhans W: Pancreatic signals controlling food intake; insulin, glucagon and amylin. Philos Trans R Soc Lond 2006;361:1219–1235.

11 Hollander P, Maggs DG, Ruggles JA, Fineman M Shen L, Kolterman OG, Weyer C: Effect of pramlintide on weight in overweight and obese insulin-treated type 2 diabetes patients. Obes Res 2004;12:661–668.

12 Kojima M, Hosoda H, Date Y, Nakazato M, Matsuo H, Kangawa K: Ghrelin is a growth-hormone releasing acylated peptide from the stomach. Nature 1999;402:656–660.

13 Wren AM, Seal LJ, Cohen MA, Brynes AE, Frost GS, Murphy KG, Dhillo WS, Ghatei MA, Bloom SR: Ghrelin enhances appetite and increases food intake in humans. J Clin Endocrinol Metab 2001;86:5992.

14 Neary NM, Small CJ, Druce MR, Park AJ, Ellis SM, Semjonous NM, Dakin CL, Filipsson K, Wang F, Kent AS, Frost GS, Ghatei MA, Bloom SR: Peptide YY3-36 and glucagon-like peptide-17-36 inhibit food intake additively. Endocrinology 2005;146:5120–5127.

15 Neary NM, Small CJ, Wren AM, Lee JL, Druce MR, Palmieri C, Frost GS, Ghatei MA, Coombes RC, Bloom SR: Ghrelin increases energy intake in cancer patients with impaired appetite: acute, randomized, placebo-controlled trial. J Clin Endocrinol Metab 2004;89:2832–2836.

16 Nagaya N, Itoh T, Murakami S, Oya H, Uematsu M, Miyatake K, Kangawa K: Treatment of cachexia with ghrelin in patients with COPD. Chest 2005;128:1187–1193.

17 Nagaya N, Moriya J, Yasumura Y, Uematsu M, Ono F, Shimizu W, Ueno K, Kitakaze M, Miyatake K, Kangawa K: Effects of ghrelin administration on left ventricular function, exercise capacity, and muscle wasting in patients with chronic heart failure. Circulation 2004;110:3674–3679.

18 Wynne K, Giannitsopoulou K, Small CJ, Patterson M, Frost G, Ghatei MA, Brown EA, Bloom SR, Choi P: Subcutaneous ghrelin enhances peritoneal dialysis: a randomized, placebo-controlled trial. J Am Soc Nephrol 2005;16:2111–2118.

19 Beck B, Richy S, Stricker-Konrad A: Feeding response to ghrelin agonist and antagonist in lean and obese Zucker rats. Life Sci 2004;76:473–478.

20 Yang J, Brown MS, Liang G, Grishin NV, Goldstein JL: Identification of the acyltransferase that octanoylates ghrelin, an appetite-stimulating peptide hormone. Cell 2008;132:387–396.

21 Gil-Campos M, Aguilera CM, Cañete R, Gil A: Ghrelin: a hormone regulating food intake and energy homeostasis. Br J Nutr 2006;96:201–226.

22 Schellekens H, Dinan TG, Cryan JF: Lean mean fat reducing 'ghrelin' machine: Hypothalamic ghrelin and ghrelin receptors as therapeutic targets in obesity. Neuropharmacology 2009 Jun 30 [Epub ahead of print].

23 Cummings DE: Ghrelin and the short- and long-term regulation of appetite and body weight. Physiol Behav 2006;89:71–84.

24 Soares JB, Roncon-Albuquerque R Jr, Leite-Moreira A: Ghrelin and ghrelin receptor inhibitors: agents in the treatment of obesity. Expert Opin Ther Targets 2008;12:1177–1189.

25 Nagamine J, Kawamura T, Tokunaga T, Hume WE, Nagata R, Nakagawa T, Taiji M: Synthesis and pharmacological profile of an orally-active growth hormone secretagogue, SM-130686. Comb Chem High Throughput Screen 2006;9:187–196.

26 Piccoli F, Degen L, MacLean C, Peter S, Baselgia L, Larsen F, Beglinger C, Drewe J: Pharmacokinetics and pharmacodynamic effects of an oral ghrelin agonist in healthy subjects. J Clin Endocrinol Metab 2007;92:1814–1820.

27 Houck WV, Pan LC, Kribbs SB, Clair MJ, McDaniel GM, Krombach RS, Merritt WM, Pirie C, Iannini JP, Mukherjee R, Spinale FG: Effects of growth hormone supplementation on left ventricular morphology and myocyte function with the development of congestive heart failure. Circulation 1999;100:2003–2009.

28 Adamopoulos S, Parissis JT, Paraskevaidis I, Karatzas D, Livanis E, Georgiadis M, Karavolias G, Mitropoulos D, Degiannis D, Kremastinos DT: Effects of growth hormone on circulating cytokine network, and left ventricular contractile performance and geometry in patients with idiopathic dilated cardiomyopathy. Eur Heart J 2003;24:2186–2196.

29 Levin F, Edholm T, Schmidt PT, Gryback P, Jacobsson H, Degerblad M, Hoybye C, Holst JJ, Rehfeld JF, Hellstrom PM, Naslund E: Ghrelin stimulates gastric emptying and hunger in normal-weight humans. J Clin Endocrinol Metab 2006;91:3296–3302.

30 Holst JJ: The physiology of glucagon-like peptide 1. Physiol Rev 2007;87:1409–1439.

31 Turton MD, O'Shea D, Gunn I, Beak SA, Edwards CM, Meeran K, Choi SJ, Taylor GM, Heath MM, Lambert PD, Wilding JP, Smith DM, Ghatei MA, Herbert J, Bloom SR: A role for glucagon-like peptide-1 in the central regulation of feeding. Nature 1996;379:69–72.

32 Abbott CR, Monteiro M, Small CJ, Sajedi A, Smith KL, Parkinson JR, Ghatei MA, Bloom SR: The inhibitory effects of peripheral administration of peptide YY(3-36) and glucagon-like peptide-1 on food intake are attenuated by ablation of the vagal-brainstem-hypothalamic pathway. Brain Res 2005;10044:127–131.

33 Baggio LL, Huang Q, Brown TJ, Drucker DJ: Oxyntomodulin and glucagon-like peptide-1 differentially regulate murine food intake and energy expenditure. Gastroenterology 2004;127:546–558.

34 Gutzwiller JP, Drewe J, Göke B, Schmidt H, Rohrer B, Lareida J, Beglinger C: Glucagon-like peptide-1 promotes satiety and reduces food intake in patients with diabetes mellitus type 2. Am J Physiol 1999;276: R1541–R1544.

35 Gutzwiller JP, Göke B, Drewe J, Hildebrand P, Ketterer S, Handschin D, Winterhalder R, Conen D, Beglinger C: Glucagon-like peptide-1: a potent regulator of food intake in humans. Gut 1999;44:81–86.

36 Verdich C, Toubro S, Buemann B, Lysgard MJ, Juul HJ, Astrup A: The role of postprandial releases of insulin and incretin hormones in meal-induced satiety: effect of obesity and weight reduction. Int J Obes Rel Metab Disord 2001;25:1206–1214.

37 Zander M, Madsbad S, Madsen JL, Holst JJ: Effect of 6-week course of glucagon-like peptide 1 on glycemic control, insulin sensitivity and beta-cell function in type 2 diabetes: a parallel-group study. Lancet 2002;359:824–830.

38 Verdich C, Flint A, Gutzwiller JP, Naslund E, Beglinger C, Hellstrom PM, Long SJ, Morgan LM, Holst JJ, Astrup A: A meta-analysis of the effect of glucagon-like peptide 1 (7–36) amide on ad libitum energy intake in humans. J Clin Endocrinol Metab 2001;86:4382–4389.

39 Schmitz O: The GLP-1 concept in the treatment of type 2 diabetes–still standing at the gate of dawn? J Clin Endocrinol Metab 2008;93:375–377.

40 Buse JB, Rosenstock J, Sesti G, Schmidt WE, Montanya E, Brett JH, Zychma M, Blonde L, LEAD-6 Study Group: Liraglutide once a day versus exenatide twice a day for type 2 diabetes: a 26-week randomized, parallel-group, multinational, open-label trial (LEAD-6). Lancet 2009;374:39–47.

41 Fineman MS, Shen LZ, Taylor K, Kim DD, Baron AD: Effectiveness of progressive dose-escalation of exenatide (exendin-4) in reducing dose-limiting side effects in subjects with type 2 diabetes. Diabete Metab Res Rev 2004;20:411–417.

42 Drucker DJ, Nauck MA: The incretin system: glucagon-like peptide-1 receptor agonists and dipeptidyl peptidase-4 inhibitors in type 2 diabetes. Lancet 2006;368:1696–1705.

43 Adrian TE, Ferri GL, Bacarese-Hamilton AJ, Fuessl HS, Polak JM, Bloom SR: Human distribution and release of a putative new gut hormone, peptide YY. Gastroenterology 1985;89:1070–1077.

44 Eberlein GA, Eysselein VE, Schaeffer M: A new molecular form of PYY: structural characterization of human PYY(3–36) and PYY(1–36). Peptides 1989; 10:797–803.

45 Grandt D, Dahms P, Schimiczek M, Eysselein VE, Reeve Jr JR, Mentlein R: Proteolytic processing by dipeptidyl aminopeptidase IV generates receptor selectivity for peptide YY (PYY). Med Klin (Munich) 1993;88:143–145.

46 Berglund MM, Hipskind PA, Gehlert DR: Recent developments in our understanding of the physiological role of PP-fold peptide receptor subtypes. Exp Biol Med (Maywood) 2003;228:217–244.

47 Batterham RL, Cowley MA, Small CJ, Herzog H, Cohen MA, Dakin CL, Wren AM, Brynes AE, Low MJ, Ghatei MA, Cone RD, Bloom SR: Gut hormone PYY(3–36) physiologically inhibits food intake. Nature 2002;418:650–654.

48 Abbott CR, Monteiro M, Small CJ, Sajedi A, Smith KL, Parkinson JR, Ghatei MA, Bloom SR: The inhibitory effects of peripheral administration of peptide YY(3–36) and glucagon-like peptide-1 on food intake are attenuated by ablation of the vagal-brainstem-hypothalamic pathway. Brain Res 2005;1044: 127–131.

49 Batterham RL, Cohen MA, Ellis SM, Le Roux CW, Withers DJ, Frost GS, Ghatei MA, Bloom SR: Inhibition of food intake in obese subjects by peptide YY3–36. N Engl J Med 2003;349:941–948.

50 Batterham RL, Ffytche DH, Rosenthal JM, Zelaya FO, Barker GJ, Withers DJ, Williams SC: PYY modulation of cortical and hypothalamic brain areas predicts feeding behaviour in humans. Nature 2007; 450:106–109.

51 Le Roux CW, Batterham RL, Aylwin SJ, Patterson M, Borg CM, Wynne KJ, Kent A, Vincent RP, Gardiner J, Ghatei MA, Bloom SR: Attenuated peptide YY release in obese subjects is associated with reduced satiety. Endocrinology 2006;147:3–8.

52 Sloth B, Holst JJ, Flint A, Gregersen NT, Astrup A: Effects of PYY1–36 and PYY3–36 on appetite, energy intake, energy expenditure, glucose and fat metabolism in obese and lean subjects. Am J Physiol Endocrinol Metab 2007;292:E1062–1068.

53 Degen L, Oesch S, Casanova M, Graf S, Ketterer S, Drewe J, Beglinger C: Effect of peptide YY-3–36 on food intake in humans. Gastroenterology 2005;129: 1430–1436.

54 Vrang N, Madsen AN, Tang-Christensen M, Hansen G, Larsen PJ: PYY (3–36) reduces food intake and body weight and improves insulin sensitivity in rodent models of diet-induced obesity. Am J Physiol Regul Integr Comp Physiol 2006;291:R367–R375.

55 Van den Hoek AM, Heijboer AC, Voshol PJ, Havekes LM, Romijn JA, Corssmit EP, Pijl H: Chronic PYY3–36 treatment promotes fat oxidation and ameliorates insulin resistance in C57BL6 mice. Am J Physiol Endocrinol Metab 2007;292:E238–E245.

56 Brandt G, Park A, Wynne K, Sileno A, Jazrawi R, Woods A, Quay S, Bloom S: Nasal peptide YY3–36. Phase 1 dose ranging and safety studies in healthy human subjects (abstract). 86th Meeting of the Endocrine Society (ENDO 2004), New Orleans, 2004.
57 Beglinger C, Poller B, Arbit E, Ganzoni C, Gass S, Gomez-Orellana I, Drewe J: Pharmacokinetics and pharmacodynamic effects of oral GLP-1 and PYY3–36:a proof-of-concept study in healthy subjects. Clin Pharmacol Ther 2008;84:468–474.
58 Thearle M, Aronne LJ: Obesity and pharmacologic therapy. Endocrinol Metab Clin North Am 2003; 32:1005–1024.
59 Curioni C, André C: Rimonabant for overweight or obesity. Cochrane Database of Systematic Reviews 2006;issue 4:Art No CD006162.
60 Roth JD, Coffey T, Jodka CM, Maier H, Athanacio JR, Mack CM, Weyer C, Parkes DG: Combination therapy with amylin and peptide YY [3–36] in obese rodents: anorexigenic synergy and weight loss additivity. Endocrinology 2007;148:6054–6061.

Christoph Beglinger, MD
Division of Gastroenterology, University Hospital
CH–4031 Basel (Switzerland)
Tel. +41 61 265 51 75, Fax +41 61 265 53 52, E-Mail beglinger@tmr.ch

Langhans W, Geary N (eds): Frontiers in Eating and Weight Regulation.
Forum Nutr. Basel, Karger, 2010, vol 63, pp 64–74

Roles of Amylin in Satiation, Adiposity and Brain Development

Thomas A. Lutz

Institute of Veterinary Physiology, and Zürich Center for Integrative Human Physiology, University of Zürich, Zürich, Switzerland

Abstract

Amylin plays an important role in the control of nutrient fluxes. It is cosecreted with insulin and reduces eating by promoting meal-ending satiation. This effect seems to depend on a stimulation of amylin receptors in the area postrema. Subsequent to area postrema activation, the neural signal is conveyed to the forebrain via distinct relays in the nucleus of the solitary tract and the lateral parabrachial nucleus to the lateral hypothalamic area and other hypothalamic nuclei; the functional roles of these relays in amylin's eating inhibitory effect have not been fully investigated. Amylin may also play a role in the regulation of adiposity. Plasma levels of amylin are increased in adiposity, although the precise relation is unknown. Furthermore, chronic infusion of amylin into the brain reduced body weight gain and adiposity, and chronic infusion of an amylin receptor antagonist increased body adiposity. Both these animal data and pre-clinical research in humans indicate that amylin is a promising option for anti-obesity therapy, especially in combination with leptin. Finally, recent findings indicate that amylin may also be necessary for normal brain development; it acts as a neurotrophic factor for the development of brainstem pathways involved in the control of eating. How this may be relevant under physiological conditions requires further studies, but these findings substantiate the concept that amylin plays an integrative role in the development and operation of neural circuits involved in the control of eating and energy homeostasis.

The best investigated function of amylin is its role as a physiological signal of satiation [1, 2]. Acute amylin injection decreases eating in rats within few minutes; amylin reduces meal size without producing a conditioned taste aversion or without increasing kaolin consumption [3, 4], and administration of the amylin receptor antagonist AC187 at a dose that blocks the eating-inhibitory effect of exogenous amylin stimulates eating by increasing meal size [5, 6]. The satiating effect of peripheral amylin seems to be mediated by action on area postrema (AP) neurons: the effect on eating is abolished in rats with specific AP lesions, but not by subdiaphragmatic vagotomy or by capsaicin-induced lesions of peripheral neural afferents projecting to the brain; local injection of amylin into the AP inhibits eating by reducing meal size; injection of the amylin receptor antagonist AC187 has the opposite effect; and electrophysiological and immunohistochemical studies confirmed a direct influence of amylin on the AP [2, 6, 7, 8].

This review focuses on four current areas of research on amylin's roles in eating and energy homeostasis. First, I discuss the central pathways involved in processing amylin signaling. Amylin activates distinct brain areas. Interestingly, there is a large overlap in these areas between different satiating signals despite differences at the behavioral level. How the brain differentiates among such apparently similar signals is not yet known. Second, I review the potential role of amylin as adiposity signal in addition to its satiating action. Third I discuss emerging data suggesting a role for amylin in brain development. This trophic effect of amylin suggests provocative questions: Do we eat to stimulate the amylin release necessary to allow our brains to develop normally? Is amylin a link between the metabolic situation and brain development in utero or in early postnatal life? These diverse actions suggest that amylin may exemplify, at least heuristically, how integrative and complex approaches are necessary to fully understand the physiological roles of signals controlling eating. Finally, I discuss the therapeutic potential of amylin in obesity.

Neural Pathways beyond the AP Processing Amylin Signaling

Amylin-induced activation of AP neurons is synaptically transmitted to the forebrain via the nucleus of the solitary tract (NTS) and the lateral parabrachial nucleus (lPBN) [8]. Techniques that have been used to define these pathways include site-specific brain lesions, c-Fos immunocytochemistry as marker of neuronal activation and retrograde and antergrade neuronal tracing studies [8–11]. Lesions of the AP or the lPBN blocked the eating-inhibitory effect of peripheral amylin and amylin-induced c-Fos expression in areas rostral to the site of lesion, i.e. in the NTS, lPBN and central nucleus of the amygdala (CeA) in AP-lesioned and in the CeA in lPBN-lesioned rats. Neuronal tracing studies confirmed direct links between these amylin-activated areas. Specifically, the lPBN appears to act as an important relay station between the hindbrain and the LHA, where amylin reduces fasting-induced c-Fos expression [8, 11]. Ascending projections to other hypothalamic nuclei, such as the ventromedial hypothalamic nucleus (VMH; discussed in Section 5.1 below) [11, 12], further processing of amylin signalling in other forebrain areas, and the circuitry eventually leading to the motor control of eating all still require investigation.

A number of peripheral peptides that inhibit eating, including amylin, cholecystokinin (CCK), glucagon-like peptide 1, and peptide YY 3–36, all produce relatively similar activation patterns in the brain [2, 8, 10, 13]. How does the brain distinguish between these different inputs? Answers to this question will probably emerge as increasingly sophisticated methods are applied. At present, the most commonly used technique to study brain activation is c-Fos immunocytochemistry after single, acute peptide administration. This defines a general activation pattern. However, one must realize that such patterns mirror only one specific time point, that they depict neuronal activation but not neuronal inhibition, that c-Fos is expressed well after the behavior of interest has occurred, and that the functional implications of c-Fos expression are unknown. Other markers now available may circumvent at least the latter two problems. For example, phosphorylated extracellular signal-regulated kinase 1/2 (pERK) in the NTS appears to be necessary for the satiating effect of CCK [14], and possibly also of amylin [unpublished]. Alternative strategies, such as extensive in vivo imaging, perhaps using functional magnetic resonance, may also help to delineate more specifically brain pathways, including a temporal pattern, for each signal. Finally, neurochemical phenotyping of neurons and tract tracing studies, as discussed, may offer strategies to extend c-Fos staining.

Amylin as an Adiposity Signal

Introduction

Several lines of evidence indicate that amylin may act as an adiposity signal as well as a satiation signal. First, amylin is considered a plausible adiposity signal because basal plasma levels of amylin are increased in obese versus lean rats [15]. Of note, lean and obese rats in that study markedly differed in age, but age per se does not seem to influence basal amylin [16]. Basal and glucose-stimulated plasma amylin levels are also elevated in obese humans [17, 18]. This suggests an association between body adiposity and plasma amylin, but further studies are warranted: (a) it is not yet clear if there is a correlation between basal plasma amylin and adiposity, as occur for leptin and insulin, and (b) it would be interesting to determine whether changes in body adiposity in individuals are reflected in changing amylin levels and whether such changes in adiposity and amylin follow the same temporal pattern. As pointed out elsewhere in this volume [19], such information is also lacking for the major hypothesized adiposity signals, leptin and insulin.

A second, direct line of evidence that amylin may be an adiposity signal is that chronic peripheral [4, 20] or central [21] amylin infusion decreased body weight gain specifically by reducing fat mass. Indeed, body-fat loss in amylin-treated rats was more pronounced than in pair-fed controls, indicating that the infusions increased energy expenditure as well as reducing food intake. These effects may possibly also be mediated by an action of amylin in the AP [22].

Third, and perhaps most importantly, chronic third ventricular administration of the amylin antagonist AC187 increased food intake and body adiposity, although body weight was unaltered [23]. Finally, the amylin knockout mouse is heavier than wildtype controls [2]. The importance of loss of amylin signaling in adulthood in these mice, however, has not been established by demonstrations that amylin replacement reverses this phenotype.

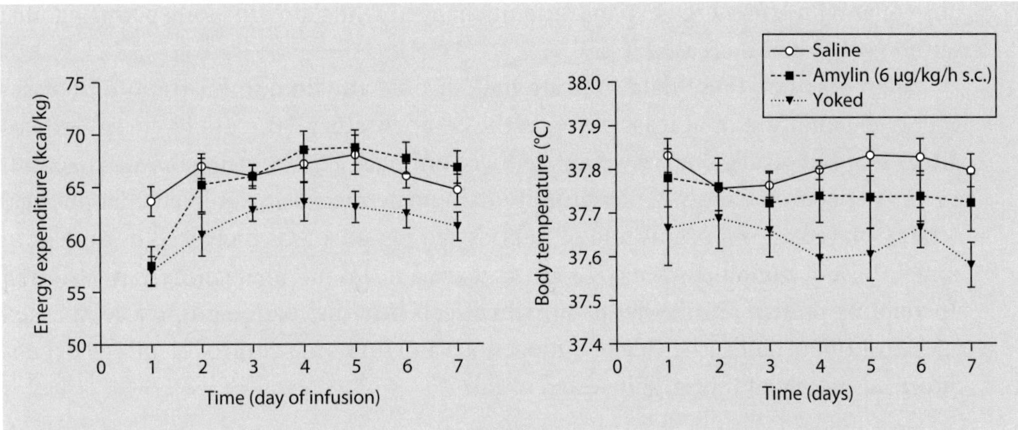

Fig. 1. Effects of chronic SC amylin infusion (6 μg/kg/h) on average dark-phase energy expenditure (**a**) and body temperature (**b**) in rats. From day 2 of infusion, amylin prevented the decrease in energy expenditure that was seen in rats whose food intake was yoked to that of amylin-treated rats. Body temperature was not affected in the amylin-treated rats, but reduced in the yoked rats. Yoked rats also received saline. n = 6 per group.

Recent experiments provided more insight into the potential role of amylin as adiposity signal [24]. Similar to what has been reported for leptin or insulin [25, 26], when tested in rats whose body weight had been previously manipulated, e.g. by two day food deprivation or by voluntary overfeeding for 3 weeks, chronic central amylin infusions led to lower body weight gain irrespective of the prior manipulation. That is, when body weight was first increased by offering an energy dense palatable food for 3 weeks, central amylin infusion reduced body weight to the same level as in amylin-infused, chow fed rats. These data suggest that, like leptin or insulin, central amylin may encode the regulated level of body weight and hence may contribute to the relative constancy of body weight throughout adult life. Of note, the specific effects on adiposity still need to be tested in this setup.

Amylin and Energy Expenditure

There are several reports that, in addition to its effect on eating, both acute and chronic amylin administration increase energy expenditure [4, 20, 27, 28]. Chronic peripheral amylin increased total energy expenditure [4, 20], which may be attributable at least in part to a relative increase in lean body mass [20]. In addition, chronic peripheral amylin infusion prevented the decrease in energy expenditure seen in controls yoke-fed to amylin-treated rats (fig. 1a). Findings were comparable during chronic central amylin infusion. The situation in regard to acute amylin administration is less clear. For example, we found that acute peripheral injection of an anorectic dose of amylin left energy expenditure unchanged while amylin's long acting agonist salmon calcitonin increased it [29].

Taken together, these data indicate that chronic amylin administration increases energy expenditure or at least prevents the compensatory decrease in energy expenditure that is typically seen in weight-reduced animals. The mechanisms of exogenous amylin's effects on energy expenditure remain unclear: it seems unlikely that physical activity played an important role [29]; body temperature was increased by amylin in some [27] but not all studies (fig. 1b), and amylin did not change the expression of uncoupling protein [20]. Whether amylin affects heat dissipation has not been tested yet. Most importantly, the role of endogenous amylin in the control of energy expenditure has not been thoroughly tested so far.

Resistance to Adiposity Signals

Many reports indicate that progressing obesity is associated with the development of resistance to the eating-inhibitory effects of exogenous leptin and insulin. Whether such resitance also occurs with amylin has not been clearly established. In the cases of leptin and insulin, obesity-related resistance appears due in part to changes in the blood-brain barrier function [30–32] (see also chapter by Banks [31]), although cellular mechanisms are also involved (see chapter by Münzberg [33]). Amylin is also transported across the blood-brain barrier [32], but it is unclear whether this transport is reduced in obesity. More importantly, however, such transport is not required for amylin action in the AP [7, 22], which is devoid of a blood-brain barrier, so that other mechanisms seem more likely to explain any potential obesity-related amylin resistance. Resistance at the cellular level (receptor or postreceptor defect) would seem more likely.

Why Distinguish between Satiation and Adiposity Signals?

The current concept of eating controls involves adiposity (tonic) signals that enhance the effect of satiation and other meal-control (episodic) signals. This concept helps to structure the multitude of signals and the complex system controlling eating; however, particular signals are not always either adiposity or meal signals. Insulin, e.g., has been hypothesized to be an adiposity signal [25, 34], but insulin is also realeased during meals and blockade of endogenous insulin by insulin antibodies increased meal size [35], indicating that insulin contributes to the control of meal termination. Similarly, ghrelin has been supposed to be involved both in meal initiation and as an (inverse) adiposity signal [34, 36].

Although amylin, like CCK, produces a satiating effect, the two differ in some respects. In particular, rats infused continuously with CCK did not show sustained reductions in food intake and body weight [37], and rats intraperitoneally infused with CCK prior to each spontaneous meal reduced meal size throughout the test, but reduced total food intake and body weight only initially because of a compensatory increase in meal frequency [38]. What exactly caused this compensatory increase is unknown. The point here is that no such compensation occurs with chronic amylin infusions despite the fact that the main direct behavioral effect in both cases is a sustained reduction in meal size [22, 39]. It is tempting to speculate that the decrease in meal size is due to a 'satiating' neural action of amylin, whereas the lack of compensatory changes in meal frequency is due to a separate 'adiposity' neural action.

As discussed above, adiposity signals appear to affect eating by enhancing the effects of satiation signals. For example, amylin, similar to leptin and insulin, enhances CCK's satiating effect [40–43]. As amylin appears to function as both an adiposity and a satiating signal, one may ask whether tonic, adiposity-signalling levels of amylin affect the satiating effect of phasic changes in amylin levels. Data are scarce. Because adult amylin-deficient mice eat less after acute peripheral amylin [unpubl.], at least the acute effect of amylin to reduce eating does not require pre-existing endogenous amylin. It remains to be tested, however, whether high tonic amylin levels increase the episodic effect of amylin on eating.

More generally, one may ask if the distinction between adiposity versus meal-control signals is useful in light of the current understanding of the dual actions of amylin, insulin [25, 34, 35] and, perhaps, ghrelin [34, 36]. The distinction would seem useful if the brain processed the two types of information differently. For example, if the two types of effects were mediated at different sites, it could be that lesioning one site would prevent one effect, but not the other. Both the acute satiating effect and the chronic adiposity effect of amylin on eating seem to require an intact AP [7, 22]. Whether subsequent processing of the two signals depends on different brain areas has not been tested. It seems that the utility of the concept of adiposity versus meal-control signals depends on evidence of distinct neural processing. Alternatively, one may consider creating a third category of signals that have both types of actions. Knowledge of the neuronal processing of all these signals, however, is not yet far enough advanced to make a final decision on this possibility.

Amylin Is Necessary for Normal Neural Development of the Brainstem

Amylin, again like leptin, has a trophic effect in neural development. Seminal studies by Bouret and colleagues (see chapter by Bouret [44]) showed that the neonatal rodent brain lacks some of the neural pathways critical for the control of energy balance. That is, genetically leptin-deficient $Lep^{ob/ob}$ mice and leptin-resistant diet-induced obese rats lack a prominent neuronal intra-hypothalalmic projection from the arcuate nucleus to the paraventricular nucleus [45, 46]. In the $Lep^{ob/ob}$ mice, early postnatal leptin treatment, which may mimic the postnatal leptin peak normally occurring between postnatal days (PN) days 4–12, restored normal neuroanatomy.

Because amylin exerts trophic effects in several peripheral tissues (e.g. kidney, bone and pancreas) [47], we tested whether amylin also acts as a trophic factor in the brain in the early postnatal period. Amylin indeed appeared to be necessary for the development of AP-NTS projections: genetically amylin-deficient mice tested on PN 10 had a markedly reduced density of AP-NTS projections compared with controls [48]. This finding opens several important questions: What is the critical period for amylin to affect AP-NTS projections? Is it possible to restore normal AP-NTS neuroanatomy in amylin-deficient mice by early postnatal amylin treatment, similar to leptin's effect in the hypothalamus [45]? Do these connections have a functional role for amylin effects on eating? In regard to the latter question, one possibility is that these connections play an important role in amylin's acute effect on eating only in early life, but, due to some structural or functional compensation, not later. This is suggested by the observation that adult amylin-deficient mice still eat less after peripheral amylin. Alternatively, these neural connections may not be necessary for amylin's acute anorectic effect in adulthood. Thus, another critical question is whether adult amylin-deficient mice in fact have similar disrupted neuroanatomy in the AP/NTS region. Finally, it will be interesting to know whether disturbed amylin signaling in fetal or early postnatal life contributes to a higher risk for developing metabolic disease because of disturbed controls of eating or energy metabolism (see also chapters by Bouret [44] and Grove [49]).

Therapeutic Potential of Amylin

Pharmacological Interactions of Amylin with Leptin in Animals

Recent research suggests that leptin and amylin interact functionally in the control of eating, and that this interaction might be useful in developing anti-obesity therapy. We initially reported that acute central leptin injection increased the eating-inhibitory effect of intraperitoneal amylin injection [50]. More recently, Roth et al. [51] tested 2-week peripheral infusions of amylin and leptin in rats with diet-induced obesity. Leptin doses that were effective in lean animals but that had no detectable effects on eating or body weight in obese rats were used. Amylin alone reduced eating significantly and led to a small decrease in body weight (~5%). Rats that received leptin and were pair-fed to amylin-treated rats did not lose more weight than amylin-treated rats. The leptin/amylin combination, however, decreased both eating and body weight more (~8–12% for body weight) than amylin alone. Further, body fat after leptin/amylin was lower than in rats receiving amylin or rats receiving leptin and pair-fed to amylin-treated rats. In other words, amylin infusion enhanced the sensitivity of obese rats to the catabolic effect of leptin [51]. The leptin/amylin combination also increased dark-phase energy expenditure [51].

Subsequent experiments using several dose combinations confirmed the synergistic effect of amylin and leptin on eating and adiposity; the combinations produced weight losses of up to 15% [52]. In that study, most of the body weight loss was due to reduced eating because rats pair-fed to the amylin/leptin group lost similar amounts of weight. Interestingly, body fat loss was more in the amylin/leptin treated rats than in the pair-fed controls. This was most likely because the pair-fed rats decreased energy expenditure, and this was prevented in the amylin/leptin group. During weight loss, the respiratory quotient was low in both the amylin/leptin and the pair-fed groups, indicating preferential oxidation of fat; but when body weight stabilized at lower level, the respiratory quotient increased in the pair-fed but not in the amylin/leptin rats. These metabolic effects in the amylin/leptin rats were mirrored in reduced expression of genes for hepatic lipogenesis and increased expression of genes for lipid utilization [52]. Finally, in both studies [51, 52] the synergistic effects of leptin/amylin on fat pad size were much clearer than the effects on body weight. This is consistent with the concept that the control mechanisms involving amylin (and leptin) primarily control adiposity, not body weight.

These data indicate that further studies are warranted to test how eating, adiposity and body weight develop after termination of treatment. This may help to develop strategies not only for weight loss but also for prolonged maintenance of lower body weights.

The amylin/leptin interaction may involve a direct effect of amylin on central leptin signalling. Amylin enhanced leptin signalling (as gauged by increased immunoreactivity of pSTAT3) specifically in the VMH [51]; previous studies had shown that amylin's eating-inhibitory effect is reduced by VMH administration of histamine H1 receptor antagonists, and that both amylin's and leptin's effects on eating are blunted in H1 receptor-deficient mice [12, 53]. Increased pSTAT3 in the VMH may be due to an amylin-induced upregulation of leptin receptor expression [unpubl.], but, because pSTAT3 immunoreactivity was also enhanced in the AP [51], identification of the exact site(s) and mechanism(s) of interaction between amylin and leptin require further studies. These should also investigate the potential role of histamine in the functional interaction of amylin and leptin.

It is interesting to consider how the pharmacological interactions of amylin and leptin on eating and adiposity translate back into the physiological situation. One question relates to the relative importance of amylin and leptin. One may argue that leptin is more important physiologically than amylin because of the more dramatic obesity phenotype of leptin-deficient mice than amylin-deficient mice. On the other hand, in obesity, leptin resistance prevents leptin from being effective, but amylin may overcome leptin resistance, suggesting that amylin may play a more important role in obesity than usually thought. Thus, leptin may be an effective starvation signal ([54] but see also chapter by Hillebrand and Geary [19]), but a relatively poor 'obesity' signal [54], whereas amylin, or amylin plus leptin, may be an effective adiposity signal.

Roth et al. [51] translated the preclinical work described above by testing co-administration of the amylin and leptin analogues pramlintide and metreleptin in overweight and obese humans. The results were highly encouraging. Body weight loss was about 5% during a 4-week pramlintide-only lead-in period. Continued treatment with either pramlintide or metreleptin for 20 weeks resulted in total body weight loss of about 8%, but combination treatment led to a weight loss of more than 12%, and weight was still decreasing at the end of the study. Side effects were mild and transient. Amylin/leptin combinations therefore seem to be a promising non-surgical approach to treat obesity. Of course, much more work is necessary to define the outcome of long-term treatment, the consequences of cessation of treatment, and potential long-term side effects.

Conclusion

Amylin is a physiological control of meal size. Most likely, this effect is initiated by amylin receptors in the AP that affect neural signaling in the NTS, lPBN, and other brain areas. Chronic amylin treatment decreases, and amylin antagonist treatment increases, eating and body weight gain, suggesting that amylin may also function as an adiposity signal. Amylin may have a positive neurotrophic effect in the early postnatal development of the hindbrain (other brain areas have not yet been studied). Animal and human studies suggest that amylin/leptin combination therapy may be an effective treatment for obesity.

Acknowledgement

The continued financial support of the Swiss National Science Foundation, the support of the Zurich Center of Integrative Human Physiology, the Novartis Foundation, the Olga Mayenfisch Foundation, and the Vontobel Foundation are gratefully acknowledged. The contributions of the members of my research group and of Profs. Scharrer, Langhans and Geary are gratefully acknowledged.

References

1 Young A, Denaro M: Roles of amylin in diabetes and in regulation of nutrient load. Nutrition 1998; 14:524–527.

2 Lutz TA: Amylinergic control of feeding. Physiol Behav 2006;89:465–471.

3 Lutz TA, Geary N, Szabady MM, Del Prete E, Scharrer E: Amylin decreases meal size in rats. Physiol Behav 1995;58:1197–1202.

4 Mack C, Wilson J, Athanacio J, Reynolds J, Laugero K, Guss S, Vu C, Roth J, Parkes D: Pharmacological actions of the peptide hormone amylin in the long-term regulation of food intake, food preference, and body weight. Am J Physiol 2007;293:R1855–R1863.

5 Reidelberger RD, Haver AC, Arnelo U, Smith DD, Schaffert CS, Permert J: Amylin receptor blockade stimulates food intake in rats. Am J Physiol 2004; 287:R568–R574.

6 Mollet A, Gilg S, Riediger T, Lutz TA: Infusion of the amylin antagonist AC 187 into the area postrema increases food intake in rats. Physiol Behav 2004;81:149–155.

7 Lutz TA, Senn M, Althaus J, Del Prete E, Ehrensperger F, Scharrer E: Lesion of the area postrema/nucleus of the solitary tract (AP/NTS) attenuates the anorectic effects of amylin and calcitonin gene-related peptide (CGRP) in rats. Peptides 1998;19:309–317.

8 Riediger T, Zuend D, Becskei C, Lutz TA: The anorectic hormone amylin contributes to feeding-related changes of neuronal activity in key structures of the gut-brain axis. Am J Physiol 2004;286: R114–R122.

9 Becskei C, Grabler V, Edwards GL, Riediger T, Lutz TA: Lesion of the lateral parabrachial nucleus attenuates the anorectic effect of peripheral amylin and CCK. Brain Res 2007;1162:76–84.

10 Rowland NE, Richmond RM: Area postrema and the anorectic actions of dexfenfluramine and amylin. Brain Res 1999;820:86–91.

11 Potes CS, Lutz TA, Riediger T: The central signaling pathways of the anorectic hormone amylin: a neuroanatomical study. Appetite 2008;51:391.

12 Mollet A, Meier S, Riediger T, Lutz TA: Histamine H1 receptors in the ventromedial hypothalamus mediate the anorectic action of the pancreatic hormone amylin. Peptides 2003;24:155–158.

13 Rinaman L, Verbalis JG, Stricker EM, Hoffman GE: Distribution and neurochemical phenotypes of caudal medullary neurons activated to express cFos following peripheral administration of cholecystokinin. J Comp Neurol 1993;338:475–490.

14 Sutton GM, Patterson LM, Berthoud HR: Extracellular signal-regulated kinase 1/2 signaling pathway in solitary nucleus mediates cholecystokinin- induced suppression of food intake in rats. J Neurosci 2004;24:10240–10247.

15 Pieber TR, Roitelman J, Lee Y, Luskey KL, Stein DT: Direct plasma radioimmunoassay for rat amylin-(1–37): concentrations with acquired and genetic obesity. Am J Physiol 1994;267:E156–E164.

16 Leckström A, Lundquist I, Ma Z, Westermark P: Islet amyloid polypeptide and insulin relationship in a longitudinal study of the genetically obese (ob/ob) mouse. Pancreas 1999;18:266–273.

17 Enoki S, Mitsukawa T, Takemura J, Nakazato M, Aburaya J, Toshimori H, Matsukara S: Plasma islet amyloid polypeptide levels in obesity, impaired glucose tolerance and non-insulin-dependent diabetes mellitus. Diabetes Res Clin Pract 1992;15:97–102.

18 Hanabusa T, Kubo K, Oki C, Nakano Y, Okai K, Sanke T, Nanjo K: Islet amyloid polypeptide (IAPP) secretion from islet cells and its plasma concentration in patients with non-insulin-dependent diabetes mellitus. Diabetes Res Clin Pract 1992;15:89–96.

19 Hillebrand JJG, Geary N: Do leptin and insulin signal adiposity?; in Langhans W, Geary N (eds): Frontiers in Eating and Weight Regulation. Forum Nutr. Basel, Karger, 2010, vol 63, pp 111–122.

20 Roth JD, Hughes H, Kendall E, Baron AD, Anderson CM: Antiobesity effects of the beta-cell hormone amylin in diet-induced obese rats: effects on food intake, body weight, composition, energy expenditure, and gene expression. Endocrinology 2006;147: 5855–5864.

21 Rushing PA, Hagan MM, Seeley RJ, Lutz TA, Woods SC: Amylin: a novel action in the brain to reduce body weight. Endocrinology 2000;141:850–853.

22 Lutz TA, Mollet A, Rushing PA, Riediger T, Scharrer E: The anorectic effect of a chronic peripheral infusion of amylin is abolished in area postrema/nucleus of the solitary tract (AP/NTS) lesioned rats. Int J Obes Relat Metab Disord 2001;25:1005–1011.

23 Rushing PA, Hagan MM, Seeley RJ, Lutz TA, D'Alessio DA, Air EL, Woods SC: Inhibition of central amylin signaling increases food intake and body adiposity in rats. Endocrinology 2001;142:5035–5038.

24 Wielinga PY, Muff S, Alder B, Woods SC, Lutz TA: Amylin levels in the brain influence the level of body weight maintenance implying that amylin acts as adiposity signal. Int J Obes 2008;32:S51.

25 Woods SC: Signals that influence food intake and body weight. Physiol Behav 2005;86:709–716.

26 Chavez M, Kaiyala K, Madden LJ, Schwartz MW, Woods SC: Intraventricular insulin and the level of maintained body weight in rats. Behav Neurosci 1995;109:528–531.

27 Osaka T, Tsukamoto A, Koyama Y, Inoue S: Central and peripheral administration of amylin induces energy expenditure in anesthetized rats. Peptides 2008;29:1028–1035.

28 Isaksson B, Wang F, Permert J, Olsson M, Fruin B, Herrington MK, Enochsson L, Erlanson-Albertsson C, Arnelo U: Chronically administered islet amyloid polypeptide in rats serves as an adiposity inhibitor and regulates energy homeostasis. Pancretology 2005;5:29–36

29 Wielinga PY, Alder B, Lutz TA: The acute effect of amylin and salmon calcitonin on energy expenditure. Physiol Behav 2007;91:212–217.

30 Banks WA: Blood-brain barrier: connecting the gut and the brain. Regul Pept 2008;149:11–14.

31 Banks WA: Blood-brain barrier as a regulatory interface; in Langhans W, Geary N (eds): Frontiers in Eating and Weight Regulation. Forum Nutr. Basel, Karger, 2010, vol 63, pp 102–110.

32 Banks WA, Kastin AJ: Differential permeability of the blood-brain barrier to two pancreatic peptides: insulin and amylin. Peptides 1998;19:883–889.

33 Münzberg H: Leptin-signaling pathways and leptin resistance; in Langhans W, Geary N (eds): Frontiers in Eating and Weight Regulation. Forum Nutr. Basel, Karger, 2010, vol 63, pp 123–132.
34 Woods SC: Gastrointestinal satiety signals. I. An overview of gastrointestinal signals that influence food intake. Am J Physiol 2004;286:G7–G13.
35 Surina-Baumgartner DM, Langhans W, Geary N: Hepatic portal insulin antibody infusion increases, but insulin does not alter, spontaneous meal size in rats. Am J Physiol 1995;269:R978–R982.
36 Williams DL, Cummings DE: Regulation of ghrelin in physiologic and pathophysiologic states. J Nutr 2005;135:1320–1325.
37 Crawley JN, Beinfeld MC: Rapid development of tolerance to the behavioural actions of cholecystokinin. Nature 1983;302:703–706.
38 West DB, Fey D, Woods SC: Cholecystokinin persistently suppresses meal size but not food intake in free-feeding rats. Am J Physiol 1984;246:R776-R787.
39 Arnelo U, Permert J, Adrian TE, Larsson J, Westermark P, Reidelberger RD: Chronic infusion of IAPP causes anorexia in rats. Am J Physiol 1996; 271:R1654–R1659.
40 Bhavsar S, Watkins J, Young A: Synergy between amylin and cholecystokinin for inhibition of food intake in mice. Physiol Behav 1998;64:557–561.
41 Riedy CA, Chavez M, Figlewicz DP, Woods SC: Central insulin enhances sensitivity to cholecystokinin. Physiol Behav 1995;58:755–760.
42 Barrachina MD, Martinez V, Wang LX, Wei JY, Tachè Y: Synergistic interaction between leptin and cholecystokinin to reduce short-term food intake in lean mice. Proc Natl Acad Sci USA 1997;94:10455–10460.
43 Mollet A, Meier S, Grabler V, Gilg S, Scharrer E, Lutz TA: Endogenous amylin contributes to the anorectic effects of cholecystokinin and bombesin. Peptides 2003;24:91–98.
44 Bouret SG: Development of hypothalamic neural networks controlling appetite; in Langhans W, Geary N (eds): Frontiers in Eating and Weight Regulation. Forum Nutr. Basel, Karger, 2010, vol 63, pp 84–93.
45 Bouret SG, Draper SJ, Simerly RB: Trophic action of leptin on hypothalamic neurons that regulate feeding. Science 2004;304:108–110.
46 Bouret SG, Gorski JN, Patterson CM, Chen S, Levin BE, Simerly RB: Hypothalamic neural projections are premanently disrupted in diet-induced obese rats. Cell Metab 2008;7:179–185.
47 Wookey PJ, Lutz TA, Andrikopoulos S: Amylin in the periphery II. Sci World J 2007;6:1642–1655.
48 Riediger T, Hermann A, Hehl A, Bouret S, Lutz TA: Amylin deficient mice have decreased fiber density in AP-NTS projections. Proceedings Society for the Study of Ingestive Behavior SSIB, 2009.
49 Sullivan EL, Grove KL: Metabolic imprinting in obesity; in Langhans W, Geary N (eds): Frontiers in Eating and Weight Regulation. Forum Nutr. Basel, Karger, 2010, vol 63, pp 186–194.
50 Osto M, Wielinga PY, Alder B, Walser N, Lutz TA: Modulation of the satiating effect of amylin by central ghrelin, leptin and insulin. Physiol Behav 2007; 91:566–572.
51 Roth JD, Roland BL, Cole RL, Trevaskis JL, Weyer C, Koda JE, Anderson CM, Parkes DG, Baron AD: Leptin responsiveness restored by amylin agonism in diet-induced obesity: evidence from nonclinical and clinical studies. Proc Natl Acad Sci USA 2008; 105:7257–7262.
52 Trevaskis JL, Coffey T, Cole R, Lei C, Wittmer C, Walsh B, Weyer C, Koda J, Baron AD, Parkes DG, Roth JD: Amylin-mediated restoration of leptin responsiveness in diet-induced obesity: magnitude and mechanisms. Endocrinology 2008;149:5679–5687.
53 Mollet A, Lutz TA, Meier S, Riediger T, Rushing PA, Scharrer E: Histmine H1 receptors mediate the anorectic action of the pancreatic hormone amylin. Am J Physiol 2001;281:R1442–R1448.
54 Schwartz MW, Woods SC, Seeley RJ, Barsh GS, Baskin DG, Leibel RL: Is the energy homeostasis system inherently biased toward weight gain? Diabetes 2003;52:232–238.

Thomas A. Lutz
Institute of Veterinary Physiology, Vetsuisse Faculty University of Zürich
Winterthurerstrasse 260
CH–8057 Zürich (Switzerland)
Tel. +41 44 635 88 08, Fax +41 44 635 89 32, E-Mail tomlutz@vetphys.uzh.ch

Langhans W, Geary N (eds): Frontiers in Eating and Weight Regulation.
Forum Nutr. Basel, Karger, 2010, vol 63, pp 75–83

The Enterocyte as an Energy Flow Sensor in the Control of Eating

Wolfgang Langhans

Physiology and Behaviour Group, Institute of Food, Nutrition and Health, ETH Zürich, Schwerzenbach, Switzerland

Abstract

Fuel monitoring in the liver or hepatic portal area was historically implicated in the control of eating. According to this view, a common denominator of nutrient metabolism such as the intracellular ATP/ADP ratio was supposed to modulate eating through changes in hepatic vagal afferent signaling. More recently, this hypothesis has been questioned because hepatic parenchymal vagal afferent innervation is scarce and because experimentally induced changes in hepatic fatty acid oxidation often failed to produce changes in eating. Accumulating evidence suggests that small intestinal enterocytes rather than hepatocytes may serve as energy flow sensors in the control of eating. These recent developments are discussed here and an outline is given of the challenges of this promising new concept.

By providing fuels and essential nutrients, eating is part of the vital regulatory feedback loops that maintain homeostasis. This homeostatic function demands that eating be controlled by metabolic feedback, but the underlying mechanisms are still unclear. Attention has focused mainly on carbohydrates and fats because changes in their utilization appear to influence eating and because they are more important fuels than proteins. As discussed in other chapters in this volume [1, 2], glucose-sensing neurons in the brain are involved in glucose homeostasis and, presumably, eating control. Hypothalamic neuronal fat metabolism also affects eating [2, 3]. Neuronal energy flow supposedly controls eating through coordinate changes in the activity of AMP-kinase (AMPK) [4, 5] and the mammalian target of rapamycin (mTOR) [6], two fuel-sensitive kinases that integrate metabolic and endocrine signals.

The existence of brain fuel sensors and their homeostatic function raises the question of whether there is a role for peripheral fuel sensing in the regulation of energy balance. Several other homeostatic feedback loops include both peripheral and central monitoring of the regulated parameter and integration of all information by the higher brain centers that control the autonomic and behavioral output [7]. This hierarchical organization provides backup, increases efficiency of the regulation and, hence, minimizes fluctuations of the regulated parameter. The control of eating presumably also uses peripheral monitoring of energy flow. Where could this occur?

Does the Liver Monitor Energy Flow?

Hepatic glucose monitoring was historically implicated in the control of eating [1]. Consistent with this, glucose inhibits eating more potently after hepatic portal vein (HPV) than after jugular vein administration [8]. The glucose sensors involved are likely located in the wall of the HPV [1] rather than in the liver parenchyma. Concepts of a more general energy flow monitoring function of the liver in the control of eating (ATP/ADP ratio, e.g. [9], fatty acid oxidation, e.g. [10]) proposed that hepatocyte metabolism influences eating through changes in ATP/ADP ratio, hepatocyte membrane potential and hepatic vagal afferent activity [9, 11]. One problem with this view is that there are barely any vagal afferent fibers in the liver parenchyma [12]. It is not clear, however, how many vagal afferents are required to relay a signal, because hepatocytes are electrochemically coupled through numerous gap junctions [13]. Nevertheless, several other problems remain: (1) An electrochemically encoded metabolic signal from hepatocytes would require a consistent relation between changes in hepatocyte membrane potential, vagal afferent activity, and food intake. Such a consistent relation does not exist [14]. For instance, the fatty acid oxidation inhibitor mercaptoacetate (MA) and the fructose polymer 2,5-anhydro-mannitol (2,5-AM), which decreases hepatic ATP [15], increased both food intake and hepatic vagal branch multiunit activity [16], but MA decreased hepatocyte membrane potential [17], whereas 2,5-AM increased it [18]. (2) The interpretations of many behavioral denervation and electrophysiological data were based on the assumption that the common hepatic branch of the vagus innervates primarily the liver, which is not true. Rather, the common hepatic branch carries mainly afferent fibers from the duodenum [12, 19]. Therefore, neither behavioral phenomena observed after section of this branch of the vagus [14, 20] nor electrophysiological data obtained in multiunit recordings from it [16, 21] can be exclusively linked to the liver. (3) Several dissociations between an inhibition of hepatic fatty acid oxidation by MA and stimulation of eating [7] question the hypothesis that changes in hepatic fatty acid oxidation modulate eating. (4) The lack of a differential eating-stimulatory effect of MA after HPV or vena cava infusion [22] also questions this hypothesis because such differences helped to identify the hepatic portal area as the site of action for glucose [8], glucagon [23], CCK-33, one of the forms of endogenous endocrine CCK [24], and 2,5-AM [25]. (5) Similar to glucose, glucagon and CCK-33 may inhibit eating by a direct effect on their receptors that are expressed on HPV vagal afferents. As 2,5-AM also stimulated eating after intracerebroventricular infusion or microinjection into the ventromedial hypothalamus [26], it can directly affect neurons and might therefore also act directly on common hepatic branch vagal afferents after peripheral administration. Thus, the observed associations between changes in hepatic energy status and changes in eating [15] need not reflect causality, and the evidence for an energy flow monitoring function of hepatocytes remains weak.

Does the Intestine Monitor Energy Flow?

Enterocyte Metabolism

The eating-stimulatory effect of MA does not appear to originate in the liver but does depend on intact abdominal vagal afferents [7, 27]. This suggests that the intestine is involved. Enterocytes require large amounts of energy for nutrient absorption and have several unique features: (1) they are exposed to and utilize fuels that enter the cells from the lumen as well as from the blood; (2) independent of its origin (gut lumen or blood), glutamine is the most important fuel of enterocytes [28], accounting for about 77 and 35% of CO_2 production in the fasted and fed states, respectively. The activity of the enzyme glutaminase, which converts glutamine to glutamate, is higher in the intestine than in most other tissues. The main end product of intestinal glutamine metabolism is alanine, which enters the HPV [28, 29]. Glutamine also provides nitrogen for the synthesis of nucleotides and other N-compounds, stimulates the mitogen-activated protein kinase pathway and promotes cell proliferation in enterocytes [30].

Enterocytes also have a high capacity for glucose utilization. Hexokinase activity is high in the fed state and decreases during fasting [29]. Even when luminal glucose is available, however, complete glucose oxidation by enterocytes is low [31, 32]. A major purpose of enterocyte glycolysis is to generate C-3 compounds such as pyruvate, lactate and alanine for the liver.

Finally, enterocytes oxidize fatty acids [31, 33, 34]. Although this is of limited energetic importance in most conditions, the capacity of enterocytes to oxidize fatty acids is implicated in the adaptation to increases in dietary fat [35, 36] and in the propensity to become obese on high-fat diets [37]. Thus, while enterocytes prefer glutamine as metabolic fuel, they can metabolize glucose and fatty acids. So far, however, effects on eating of modulations of enterocyte nutrient utilization or energy status have not been studied.

We recently observed that MA acutely and potently stimulated eating after intrajejunal infusion, and that this effect required abdominal vagal afferent signaling (table 1) [Egle et al., unpubl.], as does the effect of MA after intraperitoneal injection [27]. Intrajejunal MA infusions were effective at doses that in our hands failed to stimulate eating when infused into the HPV [22], which suggests an intestinal site of action. We obtained similar results with intragastric infusion of the glucose antimetabolite 2-deoxy-D-glucose (2DG, table 1) [Schober et al., unpubl.]. Although 2DG is scarcely absorbed, these data support the idea that intestinal fuel monitoring is not limited to fatty acid oxidation.

Table 1. Intrajejunal mercaptoacetate (MA) and intragastric 2DG infusions acutely stimulate eating in Sham but not in SDA rats

Treatment	60 min food intake, g	
	Sham (n = 12)	SDA (n = 13)
Control	0.4±0.2	0.8±0.3
MA (200 µmol/kg BW)	1.6±0.4*	0.7±0.3
	Sham (n = 13)	SDA (n = 12)
Control	0.3±0.2	0.5±0.2
2DG (400 mg/kg BW)	1.9±0.4*	0.9±0.4

Separate groups of adult male rats were kept on a fat-enriched diet [10] (rats that received MA) or on ground chow (rats that received 2DG) and underwent subdiaphragmatic vagal deafferentation (SDA) or sham surgery. After recovery from surgery, infusions of MA (2.66 ml/kg BW), 2DG (2 ml/kg BW), or equivalent volumes of equiosmotic saline were given over 30 s at 3 h into the light phase in separate within-subjects crossover designs for each infusate and surgical group. The difference between treatments is greater in sham-operated than in SDA rats, *$p < 0.05$, sequentially rejective Bonferroni t test after significant ANOVA.

Intestinal Vagal Afferents

Berthoud et al. [38] identified three types of vagal afferent terminal structures in the intestinal wall: intramuscular arrays (IMAs), intraganglionic laminar endings (IGLEs), and mucosal terminals without specific structures that ended freely in the lamina propria of the mucosa near the tip of a duodenal villus. Whereas IGLEs and IMAs presumably function as mechanosensors, the mucosal terminals respond to chemical stimuli, such as serotonin, glutamate, ATP, and other substances supposedly released mainly from the enteroendocrine cells.

Given the anatomical distribution of the common hepatic branch of the vagus [12, 39], it is possible that both the increase in multiunit vagal afferent activity by MA [21] and the loss of the eating-stimulatory effect of MA after common hepatic branch vagotomy [40], which were interpreted as evidence for an hepatic action of MA, resulted from an effect of MA on duodenal vagal afferents. Using multiunit recordings from both the common hepatic and the celiac branch of the vagus, Randich et al. [41] observed that HPV infusion of 800 µmol/kg MA increased the activity of both vagal branches similarly. Note, however, that the dose of 800 µmol/kg MA used in this study is higher than the threshold doses for stimulation of eating after intraperitoneal administration. We recently observed that 200 µmol/kg MA reliably increased the activity of serotonin-sensitive celiac vagal afferent single units [42] when infused into the superior mesenteric artery, which supplies a major part of the small intestine. This result implicates the increase in celiac vagal afferent signaling in the eating-stimulatory effect of MA. A direct effect of MA on vagal afferent fibers can presently not be excluded, but appears unlikely because peripheral nerves have only a minimal capacity to oxidize fatty acids [21]. The data are therefore consistent with the idea that an intestinal fatty acid oxidation sensing mechanism influences eating. Further support for this idea is derived from the putative role of intestinal fatty acid oxidation in the resistance to high-fat diet-induced obesity (see above) and from the demonstration that sectioning the gastroduodenal branch of the vagus antagonized the streptozotocin-induced increase in lard intake in insulin-dependent diabetic rats [43].

Challenges

The idea that enterocytes serve as energy flow sensors in the control of eating raises several questions: First and foremost, how can changes in enterocyte metabolism affect vagal afferent activity? Intestinal vagal afferents terminate in the lamina propria of the mucosa, which is close to, but still separated from, enterocytes by a basal membrane [38]. This suggests that a chemical mediator released by enterocytes into the interstitial fluid in relation to intracellular energy flow modulates vagal afferent activity. Several substances could serve this function: (1) Oleoylethanolamide (OEA), the amide of ethanolamine and oleic acid, is an endogenous lipid that is synthesized by enterocytes in response to fat intake [44]. OEA inhibits eating supposedly through vagal afferent signaling [44]. (2) The volatile neurotransmitter nitric oxide (NO), which is produced in large amounts by enterocytes and may act as an eating-stimulatory signal [45]; inhibition of NO production is implicated in the vagally mediated eating-inhibitory effect of OEA [44]. (3) The excitatory neurotransmitter glutamate, which is the major derivative of glutamine, the major fuel of enterocytes. The ionotropic glutamate receptor N-methyl-D-aspartate (NMDA) is present on vagal and non-vagal intestinal afferents [46]. Glutamate increases multiunit vagal afferent activity [47], and the delayed satiation in response to the noncompetitive NMDA-receptor blocker MK-801 depended on vagal afferents [48]. Also, intraperitoneal administration of MK-801 initially blocked, and later enhanced, the eating-stimulatory effect of MA [49], suggesting that NMDA receptor activation is involved in the initial stimulation of eating by MA. Of course, OEA, NO and glutamate are only examples for possible paracrine mediators, and whether the release of any of these three substances is modulated by changes in enterocyte energy flow is unknown. OEA is linked to fat intake and therefore seems unlikely to encode enterocyte energy flow, but NO and glutamate might qualify for such a broad signaling function. Further studies should address this possibility.

The hypothesis that MA stimulates eating by increasing intestinal vagal afferent activity is difficult to reconcile with the activation of vagal afferents by CCK or gastric distension, both of which are associated with an inhibition of eating. It is unlikely that the opposite behavioral effects of CCK and MA are mediated by stimulation of separate fibers because we have observed that most celiac single units react to both MA and CCK [42]. Therefore, the different pattern of activation that MA and CCK induce, i.e., the fast, transient response to CCK vs. the delayed, long-term response to MA may encode opposite behavior, and/or the behavioral reaction (start or stop eating) to the afferent signal might depend on its integration with other, context-specific inputs (intra-meal signals for CCK vs. between meal signals for MA) in the nucleus tractus solitarii or in higher brain centers. CCK induces satiation when injected at meal onset, whereas MA triggers a meal in animals that are not eating, but usually fails to increase meal size. Also, MA usually triggered a meal within 10–15 min after administration, i.e., within the same time it increased celiac vagal afferent activity. In any case, further experiments are necessary to clarify this point.

Another important question is whether the proposed enterocyte energy flow monitoring mechanism differentiates between macronutrients or is tuned into a common measure of intracellular energy flow? AMPK and mTOR, the two kinases implicated in central nervous system fuel sensing [4–6], contribute in enterocytes to the control of absorption [50] and cell proliferation [51], respectively. Whether changes in enterocyte mTOR or AMPK activity affect eating is unknown, but IP injection of metformin, a potent activator of AMPK that usually reduces food intake presumably through its systemic metabolic effects, has been shown to acutely and transiently stimulate eating [52]. This is consistent with a possible role of intestinal AMPK in the control of eating.

Finally, it is unknown whether the proposed energy flow monitoring mechanism differentiates between absorbed and circulating fuels? There are marked changes in enterocyte substrate utilization in relation to eating [31, 32, 34] which suggest that luminal fuels are preferred if they are available. The capability to differentiate between luminal and blood fuels would be useful for an enterocyte energy flow monitoring mechanism to detect the meal-related nutrient fluctuation and to translate it into vagal afferent signals controlling eating behavior.

Conclusion

GI peptide-encoded nutrient sensing is one form of GI vagal sensing that contributes to the effect of luminal nutrients on eating [53] and metabolism [54]. An additional energy flow-sensing mechanism could provide a more direct and presumably more accurate measure of the energy available from ingested food than the indirect measure derived from GI peptide encoding [55]. To control eating efficiently, the brain needs information about the nutrient composition and energy potential of the incoming food, and it seems advantageous to obtain this information as accurately and rapidly as possible. Intestinal cells are in the ideal position to gather such information.

While enteroendocrine cells may detect the nutrient composition of a meal and prepare the body for its handling, for example by contributing to the control of insulin secretion, enterocytes appear to be well suited to monitor potential metabolic energy. They are exposed to greater fluctuations in the availability of fuels than cells in any other organ, including the brain, and the fuels that pass through the enterocytes are a good predictor of the incoming energy load. Furthermore, enterocytes see absorbed fuels earlier than any other organ. Fats in particular reach other organs only after the delay of lymphatic absorption, i.e., enterocyte recognition of dietary fat-derived energy would save significant time in providing a full energetic inventory of ingested nutrients. Last but not least, it makes physiological sense to have an energy flow monitoring mechanism in the enterocyte because nutrient absorption demands energy. In sum, the enterocyte is an attractive candidate site for peripheral energy flow sensing in the control of eating. So far mostly indirect evidence supports this hypothesis, but it deserves to be critically examined.

References

1 Langhans W, Geary N: Overview of the physiological control of eating; in Langhans W, Geary N (eds): Frontiers in Eating and Weight Regulation. Forum Nutr. Basel, Karger, 2010, vol 63, pp 9–53.

2 Moran TH: Hypothalamic nutrient sensing and energy balance; in Langhans W, Geary N (eds): Frontiers in Eating and Weight Regulation. Forum Nutr. Basel, Karger, 2010, vol 63, pp 94–101.

3 Loftus TM, Jaworsky DE, Frehywot GL, et al: Reduced food intake and body weight in mice treated with fatty acid synthase inhibitors. Science 2000;288:2379–2381.

4 Andersson U, Filipsson K, Abbott CR, et al: AMP-activated protein kinase plays a role in the control of food intake. J Biol Chem 2004;279:12005–12008.

5 Minokoshi Y, Alquier T, Furukawa N, et al: AMP-kinase regulates food intake by responding to hormonal and nutrient signals in the hypothalamus. Nature 2004;428:569–574.

6 Cota D, Proulx K, Smith KA, et al: Hypothalamic mTOR signaling regulates food intake. Science 2006;312:927–930.

7 Langhans W: Fatty acid oxidation in the energostatic control of eating: a new idea. Appetite 2008;51:446–451.

8 Tordoff MG, Friedman MI: Hepatic portal glucose infusions decrease food intake and increase food preference. Am J Physiol 1986;251:R192–R196.

9 Friedman MI: An energy sensor for control of energy intake. Proc Nutr Soc 1997;56:41–50.

10 Scharrer E, Langhans W: Control of food intake by fatty acid oxidation. Am J Physiol 1986;250:R1003–R1006.

11 Langhans W, Scharrer E: Metabolic control of eating. World Rev Nutr Diet. Basel, Karger, 1992, vol 70, pp 1–67.

12 Berthoud HR: Anatomy and function of sensory hepatic nerves. Anat Rec A Discov Mol Cell Evol Biol 2004;280:827–835.

13 Iwai M, Miyashita T, Shimazu T: Inhibition of glucose production during hepatic nerve stimulation in regenerating rat liver perfused in situ: possible involvement of gap junctions in the action of sympathetic nerves. Eur J Biochem 1991;200:69–74.

14 Langhans W: Role of the liver in the metabolic control of eating: what we know–and what we do not know. Neurosci Biobehav Rev 1996;20:145–153.

15 Ji H, Graczyk-Milbrandt G, Friedman MI: Metabolic inhibitors synergistically decrease hepatic energy status and increase food intake. Am J Physiol 2000;278:R1579–R1582.

16 Lutz TA, Niijima A, Scharrer E: Intraportal infusion of 2,5-anhydro-D-mannitol increases afferent activity in the common hepatic vagus branch. J Auton Nerv Syst 1996;61:204–208.

17 Boutellier S, Lutz TA, Volkert M, Scharrer E: 2-Mercaptoacetate, an inhibitor of fatty acid oxidation, decreases the membrane potential in rat liver in vivo. Am J Physiol 1999;277:R301–R305.

18 Scharrer E, Rossi R, Sutter DA, Seebacher MC, Boutellier S, Lutz TA: Hyperpolarization of hepatocytes by 2,5-AM: implications for hepatic control of food intake. Am J Physiol 1997;272:R874–R878.

19 Horn CC, Friedman MI: Separation of hepatic and gastrointestinal signals from the common 'hepatic' branch of the vagus. Am J Physiol 2004;287:R120–R126.

20 Langhans W: Metabolic and glucostatic control of feeding. Proc Nutr Soc 1996;55:497–515.

21 Lutz TA, Diener M, Scharrer E: Intraportal mercaptoacetate infusion increases afferent activity in the common hepatic vagus branch of the rat. Am J Physiol 1997;273:R442–R445.

22 Mansouri A, Arnold A, Geary N, Leonhardt M, Langhans W: Mercaptoacetate (MA) stimulates feeding after infusion into the hepatic portal vein (HPV) or vena cava (VC), but not after infusion into the descending aorta. Appetite 2008;51:384.

23 Geary N, Le Sauter J, Noh U: Glucagon acts in the liver to control spontaneous meal size in rats. Am J Physiol 1993;264:R116–R122.

24 Eisen S, Phillips RJ, Geary N, Baronowsky EA, Powley TL, Smith GP: Inhibitory effects on intake of cholecystokinin-8 and cholecystokinin-33 in rats with hepatic proper or common hepatic branch vagal innervation. Am J Physiol 2005;289:R456–R462.

25 Tordoff MG, Rawson N, Friedman MI: 2,5-Anhydro-D-mannitol acts in liver to initiate feeding. Am J Physiol 1991;261:R283–R288.

26 Sakata T, Kurokawa M: Feeding modulation by pentose and hexose analogues. Am J Clin Nutr 1992;55(1 suppl):272S–277S.

27 Brandt K, Arnold M, Geary N, Langhans W, Leonhardt M: Vagal afferents mediate the feeding response to mercaptoacetate but not to the beta (3) adrenergic receptor agonist CL 316,243. Neurosci Lett 2007;411:104–107.

28 Newsholme P, Procopio J, Lima MM, Pithon-Curi TC, Curi R: Glutamine and glutamate–their central role in cell metabolism and function. Cell Biochem Funct 2003;21:1–9.

29 Newsholme EA, Carrie AL: Quantitative aspects of glucose and glutamine metabolism by intestinal cells. Gut 1994;35(1 suppl):S13–S17.

30 Rhoads JM, Wu GY: Glutamine, arginine, and leucine signaling in the intestine. Amino Acids 2009;37:111–122.

31 Duee PH, Darcy-Vrillon B, Blachier F, Morel MT: Fuel selection in intestinal cells. Proc Nutr Soc 1995;54:83–94.

32 Windmueller HG, Spaeth AE: Respiratory fuels and nitrogen metabolism in vivo in small intestine of fed rats: quantitative importance of glutamine, glutamate, and aspartate. J Biol Chem 1980;255:107–112.

33 Fleming SE, Fitch MD, DeVries S, Liu ML, Kight C: Nutrient utilization by cells isolated from rat jejunum, cecum and colon. J Nutr 1991;121:869–878.

34 Storch J, Zhou YX, Lagakos WS: Metabolism of apical vs. basolateral sn-2-monoacylglycerol and fatty acids in rodent small intestine. J Lipid Res 2008;49:1762–1769.

35 Mori T, Kondo H, Hase T, Tokimitsu I, Murase T: Dietary fish oil upregulates intestinal lipid metabolism and reduces body weight gain in C57BL/6J mice. J Nutr 2007;137:2629–2634.

36 Murase T, Aoki M, Wakisaka T, Hase T, Tokimitsu I: Anti-obesity effect of dietary diacylglycerol in C57BL/6J mice: dietary diacylglycerol stimulates intestinal lipid metabolism. J Lipid Res 2002;43:1312–1319.

37 Kondo H, Minegishi Y, Komine Y, Mori T, Matsumoto I, Abe K, et al: Differential regulation of intestinal lipid metabolism-related genes in obesity-resistant A/J vs. obesity-prone C57BL/6J mice. Am J Physiol 2006;291:E1092–E1099.

38 Berthoud HR, Blackshaw LA, Brookes SJ, Grundy D: Neuroanatomy of extrinsic afferents supplying the gastrointestinal tract. Neurogastroenterol Motil 2004;16(suppl 1):28–33.

39 Berthoud HR, Neuhuber WL: Functional and chemical anatomy of the afferent vagal system. Auton Neurosci 2000;85:1–17.

40 Langhans W, Scharrer E: Evidence for a vagally mediated satiety signal derived from hepatic fatty acid oxidation. J Auton Nerv Syst 1987;18:13–18.

41 Randich A, Spraggins DS, Meller ST, Kelm GR, Cox JE: Responses of hepatic and celiac vagal afferents to intraportal mercaptoacetate in rats fed a high-fat or low-fat diet. Neuroreport 2002;13:675–679.

42 Arnold M, Langhans W: Mercaptoacetate (MA) increases intestinal vagal afferent activity. Appetite 2009;52:816.

43 Warne JP, Foster MT, Horneman HF, Pecoraro NC, de Jong HK, Ginsberg AB et al: The gastroduodenal branch of the common hepatic vagus regulates voluntary lard intake, fat deposition, and plasma metabolites in streptozotocin-diabetic rats. Am J Physiol 2008;294:E190–E200.

44 Lo VJ, Gaetani S, Fu J, Oveisi F, Burton K, Piomelli D: Regulation of food intake by oleoylethanolamide. Cell Mol Life Sci 2005;62:708–716.

45 Janero DR, Barrnett R: Cellular and thylakoid-membrane glycolipids of *Chlamydomonas reinhardtii* 137+. J Lipid Res 1981;22:1119–1125.

46 Kirchgessner AL: Glutamate in the enteric nervous system. Curr Opin Pharmacol 2001;1:591–596.
47 Niijima A: Reflex effects of oral, gastrointestinal and hepatoportal glutamate sensors on vagal nerve activity. J Nutr 2000;130(4S suppl):971S–973S.
48 Burns GA, Fleischmann LG, Ritter RC: MK-801 interferes with nutrient-related signals for satiation. Appetite 1998;30:1–12.
49 Duva MA, Siu A, Stanley BG: The NMDA receptor antagonist MK-801 alters lipoprivic eating elicited by 2-mercaptoacetate. Physiol Behav 2005;83:787–791.
50 Walker J, Jijon HB, Diaz H, Salehi P, Churchill T, Madsen KL: 5-Aminoimidazole-4-carboxamide riboside (AICAR) enhances GLUT2-dependent jejunal glucose transport: a possible role for AMPK. Biochem J 2005;385:485–491.
51 Nakajo T, Yamatsuji T, Ban H, et al: Glutamine is a key regulator for amino acid-controlled cell growth through the mTOR signaling pathway in rat intestinal epithelial cells. Biochem Biophys Res Commun 2005;326:174–180.
52 Del Prete E, Lutz TA, Scharrer E: Acute increase in food intake after intraperitoneal injection of metformin in rats. Physiol Behav 1999;67:685–689.
53 Sclafani A, Ackroff K, Schwartz GJ: Selective effects of vagal deafferentation and celiac-superior mesenteric ganglionectomy on the reinforcing and satiating action of intestinal nutrients. Physiol Behav 2003;78:285–294.
54 Wang PYT, Caspi L, Lam CKL, et al: Upper intestinal lipids trigger a gut-brain-liver axis to regulate glucose production. Nature 2008;452:1012–1016.
55 Woods SC: The control of food intake: behavioral versus molecular perspectives. Cell Metab 2009;9:489–498.

Wolfgang Langhans
Physiology and Behaviour Group, Institute of Food, Nutrition and Health, ETH Zürich
Schorenstrasse 16
CH–8603 Schwerzenbach (Switzerland)
Tel. +41 44 655 7420, Fax +41 44 655 7206, E-Mail wolfgang-langhans@ethz.ch

Langhans W, Geary N (eds): Frontiers in Eating and Weight Regulation.
Forum Nutr. Basel, Karger, 2010, vol 63, pp 84–93

Development of Hypothalamic Neural Networks Controlling Appetite

Sebastien G. Bouret[a,b]

[a]The Saban Research Institute, Neuroscience Program, Childrens Hospital Los Angeles, University of Southern California, Los Angeles, Calif., USA; [b]Inserm, Jean-Pierre Aubert Research Center, U837, University of Lille 2, Lille, France

Abstract

The hypothalamus plays an essential role in controlling appetite during adult life. It undergoes tremendous growth beginning early in gestation and continuing during the postnatal period. During this developmental period, a variety of processes shape the hypothalamic nuclei involved in the control of eating. These include the birth of new cells that populate these areas (neurogenesis), the migration of these cells to their final destinations, selective neuronal death, and, finally, the development of functional neural connections. Each of these developmental processes represents an important period of vulnerability during which alterations of the pre- (intrauterine) and early postnatal environments may have long-term and potentially irreversible consequences on hypothalamic development and function. Metabolic hormones, including the adipocyte-derived hormone leptin, have recently emerged as likely mediators of the environmental nutrient-sensing apparatus that directs hypothalamic programming.

A major outcome of brain development is production of the necessary neural architecture for integrating information from the external environment with internal cues that reflect important aspects of an animal's physiological state. This integration allows the elaboration of adaptive behavioral and physiological responses that are essential for survival. The hypothalamus plays an essential role in these functions by integrating endocrine, autonomic and somatomotor control mechanisms that coordinate a variety of neuroendocrine homeostatic processes [1]. One unique property of hypothalamic development, as compared to development of other brain structures such as the cortex and hippocampus, is that it is to a large degree activity-independent, but instead is controlled by physiological signals that reflect environmental (nutritional) conditions.

The formation of neural systems, including the hypothalamus, is characterized by three major stages. The first stage is that of neurogenesis, during which precursors cells give rise to differentiating cells. The second major stage is the migration of neurons to form the various nuclei and areas that constitute the specific brain structure in question. Finally, axons from fully differentiated neurons project to their target sites and form synapses, thereby facilitating specific functions and behaviors. In rodents, the hypothalamus develops during a relatively long period, beginning early in gestation and continuing during the postnatal period. The developing hypothalamus is therefore exposed to two different and successive environments: one intrauterine, during gestation, and the other, extrauterine, during postnatal life. These developmental windows represent periods of vulnerability for hypothalamic development during which alterations in the nutritional and/or hormonal environment of the animal may perturb normal development and, consequently, subsequent function. However, before assessing the influence of perinatal environmental factors in hypothalamic development, it is critical that we have a good understanding of the time lines of normal hypothalamic development in species that are used for study of metabolic programming, such as the mouse and the rat.

Hypothalamic Neurogenesis and Cell Migration

The process of developing a functional hypothalamic nucleus begins with the birth or terminal mitosis of neurons in that region. Much of what we know about generation of hypothalamic nuclei has been inferred from studies in rodents. In their landmark analysis of the fetal rat hypothalamus, Altman and Bayer [2] revealed that cells composing hypothalamic nuclei derive primarily from precursors originating in the proliferative zone surrounding the lower portion of the third ventricle. This proliferative zone is also known as the neuroepithelium of the third ventricle. More recent birthdating studies employing modern markers of neurogenesis, such as the thymidine analog BrdU, have allowed scientists to obtain more detailed and precise information on the generation of hypothalamic nuclei. These studies indicate that, in mice, the majority of hypothalamic neurons located in hypothalamic nuclei known to play a role in eating and body weight regulation are born between embryonic day (E) 12 and E16 [3] (fig. 1). Those that populate the dorsomedial (DMH) and the paraventricular (PVH) nuclei of the hypothalamus are born between E12 and E14. The arcuate (Arc) and ventromedial (VMH) nuclei of the hypothalamus have relatively long neurogenic periods, with the majority of neurons found in these nuclei born as early as E12, but some as late as E16. In contrast, the lateral hypothalamic area (LHA) has a short neurogenic period that is restricted to E12. The times at which specific hypothalamic cell types are generated are largely unknown. Using the BrdU method, Brischoux et al. [4] reported that, in rats, the majority of melanin-concentrating hormone (MCH) neurons in the LHA are born between E12 and E13. Interestingly, MCH mRNA is detected in the LHA as early as E13 [4]. These anatomical observations are consistent with early determination of cell fate. Gene expression studies have also shown that

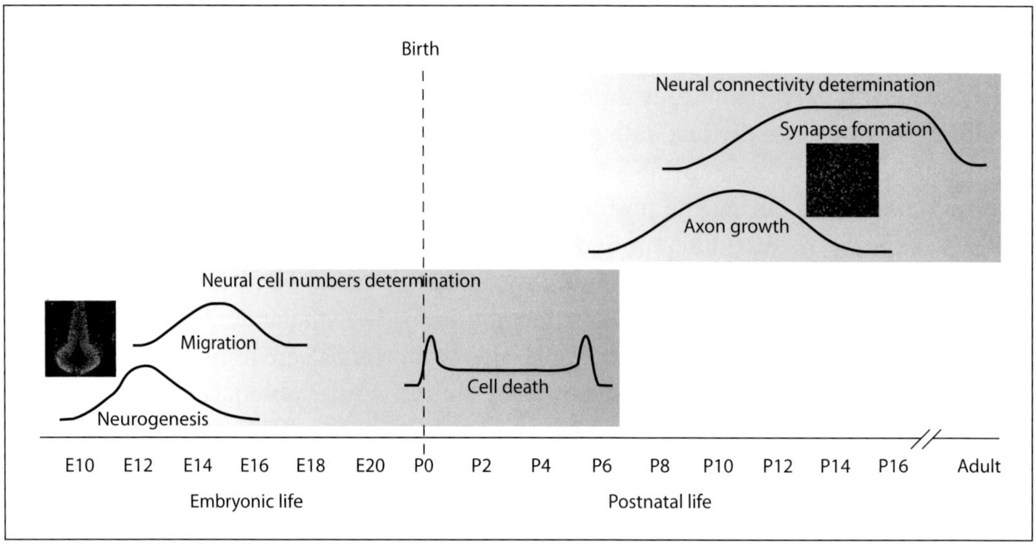

Fig. 1. Illustration of mouse hypothalamic development, showing periods of hypothalamic neuroepithelial cell proliferation (neurogenesis), cell migration, cell death, axonal growth, and synapse formation. Each of these developmental processes represents an important period of vulnerability, during which alterations in the pre- (intrauterine) and postnatal environments may have long-term and potentially irreversible consequences on hypothalamic cell number and connectivity.

neurons in the Arc first express proopiomelanocortin (POMC) mRNA on E12 [5] and that neuropeptide Y (NPY)-immunoreactive cell bodies are found in the Arc on E14 [6]. The expression of both orexigenic (NPY, AgRP) and anorexigenic (POMC, CART) neuropeptide mRNAs continues to increase in the Arc during the postnatal period, reaching maximal expression levels by postnatal day (P) 15 [7].

Part of the complex process of hypothalamic development also includes the proper migration of neurons from their sites of origin (i.e. the neuroepithelium of the third ventricle) to their final positions in the adult brain. One of the most well-characterized migration routes is that of neurons comprising the VMH [8]. This nucleus begins to appear in Nissl-stained sections as a distinct oval-shaped collection of cells on either side of the third ventricle around E16 and E17 in mice. To form the VMH, postmitotic neurons migrate laterally from the proliferative zone of the third ventricle. Radial glial fibers lining the third ventricle send long processes that extend to the pial surface of the brain. Cells migrate along this route to form the VMH. In addition, it should also be remembered that the numbers of neural and glial cells in the mature hypothalamus are functions not only of cell proliferation and migration, but also of cell death. Paradoxically, apoptosis is a principle of brain maturation; by the elimination of surplus newly generated cells, this process leads to optimal brain function. The developing hypothalamus is subject to several waves of cell death. For

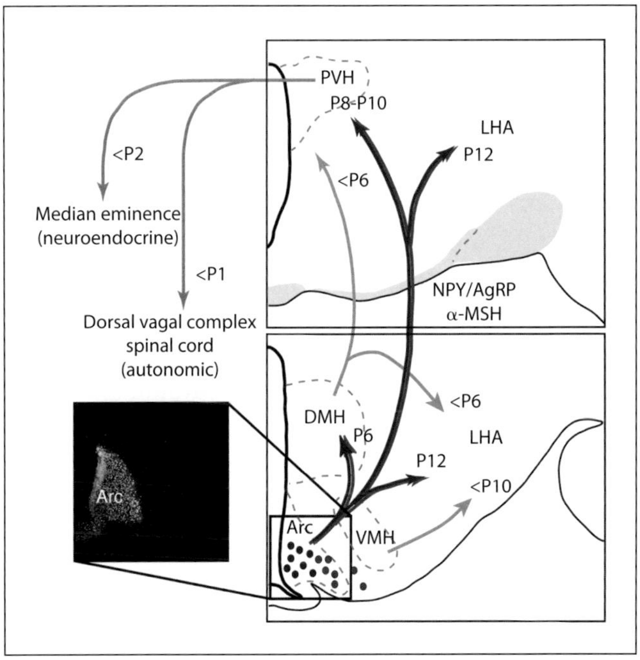

Fig. 2. Development of hypothalamic connectivity. Neural projections from the arcuate nucleus (Arc) develop postnatally and in a distinct temporal domain. Arc projections to the dorsomedial nucleus (DMH) are established by the 6th postnatal day (P6), whereas those to the paraventricular nucleus (PVH) develop significantly later, with the mature pattern of innervation first apparent between P8 and P10. Projections from the Arc to the lateral hypothalamic area (LHA) are established by P12. Leptin appears to act as a major neurotrophic signal during postnatal life to direct development of Arc projections. Interestingly, projections from the DMH, the PVH, and the ventromedial nucleus of the hypothalamus (VMH) develop prior to those from the Arc.

example, two waves of apoptosis occur in the postnatal mouse hypothalamus, one just after birth and another one at P5-P6 [Gibbons and Bouret, unpubl. data].

Development of Hypothalamic Neural Connectivity

Anatomical studies have defined the time periods during which patterns of hypothalamic connectivity are established (fig. 2). In part because of its importance for the control of eating, the first systematic study utilizing axonal labeling defined the ontogeny of projection pathways from the Arc [9]. The key finding in this study was that, although the birth and migration of Arc neurons is for the most part completed by the second half of the gestational period in rodents, the development of axonal projections from the Arc to each of its target nuclei does not occur until the second week of postnatal life. By P6, arcuate projections extend through the periventricular zone of the hypothalamus to provide inputs to the DMH, followed by inputs to the PVH between P8 and P10. Projections from the Arc to the LHA develop significantly later, with the mature pattern of innervation first apparent on P12. Not until P18 does the pattern of Arc axonal projections achieve a distribution resembling that seen in the adult.

Although the peptidergic identity of Arc projections during the postnatal period has not been fully established, NPY, AgRP, and POMC-derived peptides are expressed in these projections in adult rats and mice [10, 11]. Importantly, terminals containing AgRP/NPY are found in the DMH, PVH, and LHA of rat neonates in a pattern that coincides with the innervation of axons from the Arc [12, 13]. The fact that microinjection of NPY directly into the PVH at P2 resulted in increased milk and water intakes suggests that NPY receptors may be present and functional in the PVH before innervation of this nucleus by Arc NPY/AgRP fibers [14]. The sequential innervation of Arc targets also suggests that leptin signaling may differentially activate neurons in these target nuclei during development. Consistent with this idea, peripheral leptin injection induced c-Fos immunoreactivity (c-Fos-IR) in the Arc as early as P6, whereas leptin-induced c-Fos was not observed in the PVH before P10 or in the LHA before P16 [9]. These results indicate that development of leptin-induced c-Fos-IR in target nuclei of the Arc coincides with their innervation by Arc axons and, further, suggest that this activation by leptin may be due to transneuronal relay of the leptin signal acting on the Arc.

In contrast to the development of projections from the Arc, efferents from the DMH to the PVH and LHA are fully established by P6 [9] (fig. 2). Projections from the VMH also develop prior to those from the Arc: by P10 VMH fibers supply strong inputs to the LHA, whereas, at this age, the LHA is almost devoid of fibers from the Arc [9]. Similarly, the neurohypophyseal pathway from the PVH to the median eminence appears to be largely formed by birth [15].

The precise molecular mechanisms responsible for the formation of hypothalamic neural circuits are not well understood. In general, axon development comprises two aspects: the physical act of extension and the molecular mechanisms underlying this process. Axons grow by sending out a highly plastic and sensitive structure called a 'growth cone', which travels toward the target and trails behind it the elongating neurite. Axonal growth cones must choose a path to follow and must decide the direction to go on this path. The pathways are defined by cell-cell interactions and diffusible chemorepulsive and chemoattractive cues [16]. Interestingly, the diffusible axon guidance cues Netrin, Slit, and Semaphorins are highly expressed in the PVH during development [17; Bouret, unpubl. data]. In addition, implanting grafts derived from embryonic Arc of wild-type mice into the third ventricle of NPY knockout mice enables restoration of NPY innervation to Arc target nucleus [18]. Therefore, local directional cues produced by target nuclei appear important for proper innervation of these targets by Arc axons. The formation of hypothalamic connections also likely involves cell adhesion molecules. Consistent with this idea, Fetissov et al. [19]

reported that mice that are deficient in contactin, which is a cell adhesion molecule involved in the formation of axonal projections, and display reduced density of AgRP and αMSH neural fibers in the PVH during postnatal development.

Further work is clearly needed to define the developmental point at which the assembly of these circuits is fully established. For example, one important unanswered question is when synapses between Arc axons and PVH neurons become structurally and functionally mature. In considering synapse formation, it appears that the development of hypothalamic neural networks is an active process that, in rodents, continues well past the third week of postnatal life. For example, Arai and colleagues [20] reported a gradual increase in the number of synapses in the Arc from birth to adulthood. Ultrastructural analysis of Arc synapses in postnatal rats revealed very few axodendritic or axosomatic synapses on P5, whereas at weaning, i.e. P20, about one-half of the synapses found in adult animals were already formed. The number of Arc synapses continues to increase after weaning to reach an adult-like pattern by P45 [20]. Colmers and collaborators also reported an increase in GABAergic terminals contacting PVH parvocellular neurons between P7-P9 and P30-P40 [21]. These findings indicate that the period of vulnerability for hypothalamic development extends well beyond embryonic and early postnatal life and strongly suggest that environmental insults that occur during adolescence and early adulthood can also perturb hypothalamic organization and subsequent function.

Environmental Factors Influencing Hypothalamic Development

The array of environmental cues that can influence neural development (i.e. neurogenesis, cell migration, axonal growth, and synaptogenesis) is substantial. Hypothalamic development appears particularly highly susceptible to a variety of metabolic insults. For example, maternal obesity and undernutrition both result in altered organization of Arc neural connectivity [22, 23]. Similarly, over- or under-nutrition imposed during early postnatal life through alteration of litter size causes abnormal development of Arc neural projections to the PVH and induces a variety of alterations in hypothalamic neuronal activity [24–26]. The exact nature of the factors mediating the adverse effects of perinatal over- and undernutrition on hypothalamic development remains largely unknown. However, it is well established that circulating hormones represent important environmental signals that can act directly on the central nervous system to regulate its development and activity. Metabolic hormones appear particularly well-suited to communicate nutrient availability in the environment to the 'metabolic' networks in the hypothalamus during perinatal development. In particular, leptin has recently emerged as a key neurotrophic signal for hypothalamic development (fig. 2). During neonatal development, food intake must be maximized to support growth, yet plasma leptin levels are relatively high during both pre- and postnatal life [27, 28]. The remarkable observation that there is a dramatic surge in circulating leptin levels that is not matched by a corresponding reduction in food intake [29], led Ahima et al. [27] to suggest that leptin itself may function as a developmental cue for brain development. This hypothesis was supported by Pan et al. [30] who demonstrated the existence of active transport of leptin across the blood-brain barrier during early stages of postnatal life [31]. Analysis of hypothalamic neural connectivity in leptin-deficient ($lep^{ob/ob}$) mice provided compelling evidence that postnatal leptin promotes development of Arc neural projections [32]. The density of projections from Arc neurons to other hypothalamic sites involved in control of food intake were severely disrupted in postnatal $lep^{ob/ob}$ mice and remain diminished throughout life [32]. Both NPY/AgRP and POMC-containing neurons appeared to be affected. As reviewed by Lutz in this volume [33], another metabolic hormone, amylin, appears to play a similar role in the development of the area postrema.

Although hypothalamic circuits remain relatively plastic throughout life (another property that leptin shares with neurotrophins is the ability to cause synaptic rearrangements) [34], as with most important developmental factors, leptin appears to determine the boundaries of this plasticity during a restricted postnatal critical period. This developmental window coincides with the neonatal leptin surge. Treatment of adult $lep^{ob/ob}$ mice with leptin did not restore Arc projections to the PVH, but daily injections of leptin between P4 and P12 rescued innervation of the PVH by Arc axons [32]. The physiological relevance of postnatal leptin has been supported by several observations. Neonatal leptin treatment caused a long-term reduction in food intake in $lep^{ob/ob}$ mice [32]. In addition, inhibition of the endogenous surge of leptin in normal rats resulted in increased susceptibility to the development of diet-induced obesity during adulthood [35]. Not only the correct amplitude, but also the correct timing of the postnatal leptin surge appears to be required for normal regulation of energy homeostasis. Thus, premature leptin surges (either prenatally acquired or induced by leptin supplementation in normal newborns) induced various metabolic deficits, including higher predisposition to obesity during adult life [36].

The specific site of action for the developmental effects of leptin is unknown but appears to include direct action on Arc neurons. Leptin receptors are expressed in Arc neurons during postnatal life and induced intracellular signaling [37]. Moreover, direct application of leptin to isolated organotypic explants of the Arc induced neurite extension [32], demonstrating the trophic activity of leptin on Arc neurons. However, it cannot be excluded that axon guidance cues secreted by targets of the Arc also participate in the formation of Arc projections. Importantly, leptin receptors are expressed in the immature hypothalamus at times earlier than P10 [38] and in brain sites other than the Arc [37]. These observations suggest that the neurotrophic actions of leptin are not restricted to the postnatal development of Arc neural projections. Furthermore, other metabolic hormones, such as insulin and ghrelin, can also exert profound neurotrophic effects on the developing hypothalamus. Insulin has a well-described role in mediating the macrosomic response to maternal diabetes. The importance of insulin in hypothalamic development was shown by Toran-Allerand et al. [39], who demonstrated that exposure of isolated preoptic area explant cultures to insulin induced robust neurite elongation and proliferation. Moreover, direct application of insulin to the region of the mediobasal hypothalamus during the critical period for development of projections from the Arc to the PVH resulted in long-term perturbations in body weight regulation and glucose homeostasis [40]. These findings clearly indicate that insulin acts on the hypothalamus to affect its development and function. They also suggest that, during neonatal life, multiple peripheral signals related to energy balance may act as trophic signals to direct key developmental events in the same hypothalamic pathways that are responsible for proper regulation of energy balance in mature animals.

Conclusion

It is clear that the brain, including the hypothalamus, is critically important for the control of eating. In recent years, there has been increasing focus on the development of hypothalamic circuits and factors that may influence their development. There is now strong evidence to suggest that changes in the hormonal and nutritional milieu during critical periods of life may permanently influence the development and functional activity of the hypothalamus. This represents a key mechanism for effecting long-term changes in weight regulation in response to perinatal events. It remains to be shown how much early life experiences determine the degree and qualitative nature of the brain response to nutritional insults later in life, as well as the periods of vulnerability for human hypothalamic development, which proceeds on a timeline of months, versus days in rodents. In any events, this growing area of research promises to open new avenues for understanding perinatally acquired predispositions to eating disorders.

References

1 Swanson LW: Cerebral hemisphere regulation of motivated behavior. Brain Research 2000;886:113–164.

2 Altman J, Bayer SA: The development of the rat hypothalamus. Adv Anat Embryol Cell Biol 1986; 100:1–178.

3 Ishii Y, Gibbons MB, Bouret SG: Neurogenic actions of leptin on hypothalamic structures involved in feeding regulation. Proc 39th Annual Meet Society for Neuroscience, Chicago, 2009.

4 Brischoux F, Fellman D, Risold P-Y: Ontogenetic development of the diencephalic MCH neurons: a hypothalamic 'MCH area' hypothesis. Eur J Neurosci 2001;13:1733–1744.

5 Khachaturian H, Alessi NE, Lewis ME, Munfakh N, Fitzsimmons MD, Watson SJ: Development of hypothalamic opioid neurons: A combined immunocytochemical and [^{3}H]thymidine autoradiographic study. Neuropeptides 1985;5:477–480.

6 Kagotani Y, Hashimoto T, Tsuruo Y, Kawano H, Daikoku S, Chihara K: Development of the neuronal system containing neuropeptide Y in the rat hypothalamus. Int J Dev Neurosci 1989;7:359–374.

7 Cottrell EC, Cripps RL, Duncan JS, et al: Developmental changes in hypothalamic leptin receptor: relationship with the postnatal leptin surge and energy balance neuropeptides in the postnatal rat. Am J Physiol Regul Integr Comp Physiol 2009; 296:R631–R639.

8 McClellana KM, Parker KL, Tobet SA: Development of the ventromedial nucleus of the hypothalamus. Front Neuroendocrinol 2006;27:193–209.
9 Bouret SG, Draper SJ, Simerly RB: Formation of projection pathways from the arcuate nucleus of the hypothalamus to hypothalamic regions implicated in the neural control of feeding behavior in mice. J Neurosci 2004;24:2797–2805.
10 Bai FL, Yamano M, Shiotani Y, et al: An arcuato-paraventricular and -dorsomedial hypothalamic neuropeptide Y-containing system which lacks noradrenaline in the rat. Brain Res 1985;331:172–175.
11 Baker R, Herkenham M: Arcuate nucleus neurons that project to the hypothalamic paraventricular nucleus: neuropeptidergic identity and consequences of adrenalectomy on mRNA levels in the rat. J Comp Neurol 1995;358:518–530.
12 Grove KL, Allen S, Grayson BE, Smith MS: Postnatal development of the hypothalamic neuropeptide Y system. Neuroscience 2003;116:393–406.
13 Nilsson I, Johansen JE, Schalling M, Hokfelt T, Fetissov SO: Maturation of the hypothalamic arcuate agouti-related protein system during postnatal development in the mouse. Dev Brain Res 2005; 155:147–154.
14 Capuano CA, Leibowitz SF, Barr GA: Effect of paraventricular injection of neuropeptide Y on milk and water intake of preweanling rat. Neuropeptides 1993; 24:177–182.
15 Daikoku S, Okamura Y, Kawano H, Tsuruo Y, Maegawa M, Shibasaki T: Immunohistochemical study on the development of CRF-containing neurons in the hypothalamus of the rat. Cell Tissue Res 1984;238:539–544.
16 Tessier-Lavigne M, Goodman CS: The molecular biology of axon guidance. Science 1996;274:1123–1133.
17 Xu C, Fan C-M: Expression of Robo/Slit and Semaphorin/Plexin/Neuropilin family members in the developing hypothalamic paraventricular and supraoptic nuclei. Gene Expression Patterns 2008; 8:502–507.
18 Fetissov SO, Kuteeva E, Hokfelt T: Directional cues for arcuate NPY projections are present in the adult brain. Exp Neurol 2003;183:116–123.
19 Fetissov SO, Bergstrom U, Johansen JE, Hokfelt T, Schalling M, Ranscht B: Alterations of arcuate nucleus neuropeptidergic development in contactin-deficient mice: comparison with anorexia and food-deprived mice. Eur J Neurosci 2005;22:3217–3228.
20 Matsumoto A, Arai Y: Developmental changes in synaptic formation in the hypothalamic arcuate nucleus of female rats. Cell Tissue Res 1976;14:143–156.
21 Melnick I, Pronchuck N, Cowley MA, Grove KL, Colmers WF: Developmental switch in neuropeptide Y and melanocortin effects in the paraventricular nucleus of the hypothalamus. Neuron 2007;56: 1103–1115.
22 Kirk SL, Samuelsson A-M, Argenton M, et al: Maternal obesity induced by diet in rats permanently influences central processes regulating food intake in offspring. PLoS ONE 2009;4:e5870.
23 Delahaye F, Breton C, Risold P-Y, et al: Maternal perinatal undernutrition drastically reduces postnatal leptin surge and affects the development of arcuate nucleus proopiomelanocortin neurons in neonatal male rat pups. Endocrinology 2008;149: 470–475.
24 Bouret SG, Burt-Solorzano C, Wang C-H, Simerly RB: Impact of neonatal nutrition on development of brain metabolic circuits in mice. Proc 37th Ann Meet Society for Neuroscience, San Diego, 2007.
25 Davidowa H, Plagemann A: Different responses of ventromedial hypothalamic neurons to leptin in normal and early postnatally overfed rats. Neurosci Lett 2000;293:21–24.
26 Davidowa H, Li Y, Plagemann A: Altered responses to orexigenic (AGRP, MCH) and anorexigenic (a-MSH, CART) neuropeptides of paraventricular hypothalamic neurons in early postnatally overfed rats. Eur J Neurosci 2003;18:613–621.
27 Ahima R, Prabakaran D, Flier J: Postnatal leptin surge and regulation of circadian rhythm of leptin by feeding: implications for energy homeostasis and neuroendocrine function. J Clin Invest 1998;101: 1020–1027.
28 Udagawa J, Hashimoto R, Suzuki H, et al: The role of leptin in the development of the cerebral cortex in mouse embryos. Endocrinology 2006;147:647–658.
29 Mistry A, Swick A, Romsos D: Leptin alters metabolic rates before acquisition of its anorectic effect in developing neonatal mice. Am J Physiol 1999;277: R742–R747.
30 Pan W, Hsuchou H, Hong T, Kastin A: Developmental changes of leptin receptors in cerebral microvessels: unexpected relation to leptin transport. Endocrinology 2008;149:2798–2806.
31 Bouret SG: Crossing the border: developmental regulation of leptin transport to the brain. Endocrinology 2008;149:875–876.
32 Bouret SG, Draper SJ, Simerly RB: Trophic action of leptin on hypothalamic neurons that regulate feeding. Science 2004;304:108–110.
33 Lutz TA: The roles of amylin in satiation, adiposity and brain development; in Langhans W, Geary N (eds): Frontiers in Eating and Weight Regulation. Basel, Karger, 2009.

34 Pinto S, Roseberry AG, Liu H, et al: Rapid rewiring of arcuate nucleus feeding circuits by leptin. Science 2004;304:110–115.

35 Attig L, Solomon G, Ferezou J, et al: Early postnatal leptin blockage leads to a long-term leptin resistance and susceptibility to diet-induced obesity in rats. Int J Obes 2008;32:1153–1160.

36 Yura S, Itoh H, Sagawa N, et al: Role of premature leptin surge in obesity resulting from intrauterine undernutrition. Cell Metabolism 2005;1:371–378.

37 Caron E, Sachot C, Prevot V, Bouret SG: Distribution of leptin-sensitive cells in the postnatal and adult mouse brain. J Comp Neurol 2009;in press.

38 Carlo AS, Williams LM: Early developmental expression of leptin receptor gene and [^{125}I]leptin binding in the rat forebrain. J Chem Neuroanat 2007; 33:155–63.

39 Toran-Allerand CD, Ellis L, Pfenninger KH: Estrogen and insulin synergism in neurite growth enhancement in vitro: mediation of steroid effects by interactions with growth factors? Dev Brain Res 1988;41:87–100.

40 Plagemann A, Heidrich I, Götz F, Rohde W, Dörner G: Lifelong enhanced diabetes susceptibility and obesity after temporary intrahypothalamic hyperinsulinism during brain organization. Exp Clin Endocrinol 1992;99:91–95.

Dr. S.G. Bouret, PhD
The Saban Research Institute, Neuroscience Program,
Childrens Hospital Los Angeles, University of Southern California
4650 Sunset Boulevard, MS#135
Los Angeles, CA 90027 (USA)
Tel. +1 323 361 8743, Fax +1 323 361 1549, E-Mail sbouret@chla.usc.edu

Langhans W, Geary N (eds): Frontiers in Eating and Weight Regulation.
Forum Nutr. Basel, Karger, 2010, vol 63, pp 94–101

Hypothalamic Nutrient Sensing and Energy Balance

Timothy H. Moran

Department of Psychiatry and Behavioral Sciences, Johns Hopkins University School of Medicine, Baltimore, Md., USA

Abstract

Hypothalamic neurons have the capacity to sense and alter their activity in response to fluctuations in local nutrient concentrations. Alterations in glucose, fatty acid and amino acid concentrations have all been demonstrated to affect neuronal excitability and/or intracellular signaling pathways. The degree to which such changes in nutrient availability have the capacity to modify energy balance varies across nutrient type. The underlying mechanisms through which various nutrients affect food intake and overall energy balance involve both specific and shared neuronal substrates.

Jean Mayer [1] originally proposed the concept of nutrient sensing in the control of energy balance. He hypothesized that decreased glucose utilization represented the stimulus for meal initiation. He defined decreased glucose utilization or 'metabolic hypoglycemia' as the point at which the peripheral arteriovenous difference in blood glucose (A-V delta glucose) becomes negligible and glucose is no longer entering metabolizing cells. He suggested that this difference was detected in the brain at glucosensitive sites, in unspecified locations. A further aspect of his proposal was that increased glucose utilization in these same glucosensitive sites would lead to decreased hunger and the cessation of eating. Other hypotheses followed based on lipid or protein sensing or overall energy utilization as signals for the control of food intake. Only recently have some of the mechanisms underlying the hypothalamic actions of specific nutrients in the controls of food intake begun to be understood.

Hypothalamic Glucose Sensing

Oomura et al. [2] identified neurons within the ventromedial and lateral hypothalamus that changed their activity in response to alterations in glucose concentrations.

Thus, whereas all neurons use glucose as their primary metabolic fuel, glucose-sensing neurons use glucose as a signaling molecule to regulate their membrane potential and control their electrophysiological activity. Two populations were initially identified, and their response characteristics have now been investigated [3]. One population increases its electrophysiological activity in response to increases in glucose concentrations. These have been termed glucose-excited (GE) or glucose-responsive neurons. The other population increases activity when the local concentration of glucose is diminished. These have been termed glucose-insensitive or glucose-inhibited (GI) neurons. Although the alterations in glucose concentrations initially used to characterize these neurons were unphysiologically high, more recent work with alterations in concentration within the range that is normally found in brain with fasting and refeeding have verified that neurons can respond to physiological changes in glucose concentrations [3]. The presence of such neurons has been demonstrated in multiple hypothalamic and hindbrain sites.

Glucose-sensing neurons within the medial basal hypothalamus (ventromedial and arcuate nuclei) have been the best characterized. An important role for the ATP-activated K^+ (K_{ATP}) channel in the activity of GE neurons has been documented [4]. With increasing glucose concentration and metabolism, the K_{ATP} channel is inactivated, leading to membrane depolarization and increased neuronal activity. Since the K_{ATP} channel is ubiquitously expressed, it is only those neurons in which the channel is gated by the amount of available glucose or alterations in the ATP to ADP ratio that can use the channel to link alterations in glucose concentration to changes in electrophysiological activity. The mechanisms underlying glucose sensing in GI neurons are less well understood.

Within the hypothalamic arcuate nucleus (Arc), anorexigenic pro-opiomelanocrtin (POMC) neurons have been demonstrated to be excited by glucose and to utilize the K_{ATP} channel to sense glucose [4]. Orexigenic neuropeptide Y/agouti related peptide (NPY/AgRP)-expressing neurons have been demonstrated to be either activated or inhibited by increasing glucose concentration, but in neither case do they utilize the K_{ATP} channel to sense glucose [5]. Arc neurons also respond to adiposity signals such as leptin and insulin [6–8], suggesting that local neuronal networks within the Arc have the capacity to respond and integrate multiple metabolic signals.

Data such as these demonstrate that the brain possesses mechanisms for sensing and responding to alterations in glucose concentration. However, they do not address whether such actions modulate food intake under normal circumstances. Peripheral glucose concentrations are low prior to eating and increase with the ingestion of carbohydrate-containing meals; the issue is whether these alterations provide signals for meal initiation and termination. Early work examining the relationships between plasma glucose and feeding suggested such a causal link. For example, Stunkard and Wolfe [9] demonstrated that a peripheral glucose infusion interrupted gastric 'hunger' contractions suggesting an inhibitory role for high glucose levels. Furthermore, initial work with the glucose analog, 2-deoxy-D-Glucose (2-DG) also provided some support for Mayer's overall hypothesis. 2-DG is a substituted glucose that is taken up by the glucose transporter, phosphorylated by hexokinase, but not further metabolized, thus preventing glycolysis and depleting the cells of glucose-derived energy ('glucoprivation'). Either peripheral or central administration of 2-DG increases food intake, suggesting that glucoprivation at some sites provides a signal for eating [10]. However, early work suggested that the hypothalamus was not the site for glucoprivation-induced eating [11] and the physiological relevance of responses to 2-DG was questioned by demonstrations that the degree of glucoprivation needed to stimulate eating was well beyond those that occur spontaneously [12]. More recent work with glucoprivic eating indicates that this is an emergency response and is mediated primarily through neurons in the dorsal hindbrain [13].

Close monitoring of blood glucose concentrations around spontaneous meals revealed a 5–10% decline in blood glucose levels during the 12–5 min preceding meal initiation (in experimental animals) [14, 15] or meal requests (in human subjects) [16]. Although the basis for these declines has not been identified, it could be demonstrated that experimentally produced declines lead to meal initiation, suggesting a causal relationship [17]. However, brain glucose levels are only 10–15% of plasma levels, and whether declines in brain levels also occur prior to spontaneous meals has only recently been addressed. Simultaneous monitoring of peripheral and basal hypothalamic glucose levels and spontaneous meal patterns demonstrated that although declines in blood glucose preceded many spontaneous meals, there was no consistent relationship between hypothalamic glucose levels and spontaneous meals [18].

Manipulations aimed at altering hypothalamic glucose sensing in specific neurons have had variable effects. Although down regulation of VMH glucokinase activity affected glucoprivic feeding, it failed to affect spontaneous eating or overall energy balance [18]. Similarly, disrupting K_{ATP} channels in POMC neurons resulted in decreased glucose-stimulated α-MSH release and impaired whole-body glucose homeostasis, but did not significantly disrupt overall energy balance [19]. Thus, even though multiple mechanisms mediating hypothalamic glucose signaling and the molecular consequences of glucose induced alterations in hypothalamic neuronal activity have been identified, no direct link between this neuronal glucose sensing and the control of eating or, in particular, the initiation and termination of spontaneous meals, has been identified.

Hypothalamic Fatty Acid Sensing

Although glucose is the main metabolic substrate for neurons, neurons also can use fatty acids for their energy needs, and recent data from a variety of sources suggest that fatty acids may act at brain sites as signals of overall metabolic state.

As well as demonstrating responses to glucose in hypothalamic neurons, Oomura et al. [20] reported alterations in electrophysiological activity after the administration of free fatty acids. More recently, Wang et al. [21] have characterized responses to oleic acid in ARC neurons. Both oleic acid-excited and oleic acid-inhibited neurons were found, and the responses of some of these neurons depended upon the ambient glucose concentration. There was only minimal overlap between oleic acid-sensing and glucose-sensing neurons, and no clear relationships between the two types emerged. For example, GE neurons could be either excited or inhibited by oleic acid, as could GI neurons. Experiments examining the mechanisms through which oleic acids affect the electrophysiological activity of VMH neurons have demonstrated multiple pathways [22]. Pharmacological antagonism of fatty acid metabolism and blockade of the fatty acid transporter CD36 both reduced, but did not fully block, the ability of oleic acid to affect VMH neuronal activity. Together, these data demonstrate the potential for fatty acids to affect hypothalamic signaling and for the effects to be dependent on metabolic status. However, a clear picture for the nature of such interactions or their underlying mechanisms has yet to emerge.

Evidence for a role for hypothalamic fatty acids in the control of eating initially derived from experiments examining the actions of a fatty acid synthase inhibitor on food intake and energy balance. As shown in figure 1, the fatty acid synthetic pathway involves the conversion of acetyl CoA to long chain fatty acid CoAs, actions catalyzed by acetyl CoA carboxylase (ACC) and fatty acid synthase (FAS). Using a FAS inhibitor originally designed as an anticancer agent (C75), Loftus et al. [23] demonstrated that peripheral or central inhibition of FAS significantly inhibited eating. C75 also prevented fasting-induced increases in hypothalamic NPY expression [23] and blocked fasting induced increases in c-Fos activation in the medial basal hypothalamus [24], suggesting a basal hypothalamic site of action. Consistent with this interpretation, FAS has been shown to be highly expressed in the basal hypothalamus and is colocalized with NPY in the ARC [25].

The initial interpretation from these data was that the increase in malonyl-CoA concentration derived from the blockade of FAS provided a signal for inhibiting eating [23]. C75 has also been shown to be a CPT-1 activator, and increased C75-induced β oxidation has also been proposed to play a signaling role for eating inhibition [26–28]. However, there is some controversy over such an action for C75 [29]. An alternative mechanism of action involves the modulation of AMP-activated protein kinase (AMPK) and cellular energy balance [30]. C75 administration has been shown to decrease pAMPK, and an AMPK activator blocked the eating-inhibitory action of C75.

Obici and colleagues [31] demonstrated that intracerebroventricular administration of oleic acid can affect eating and glucose homeostasis. Intracerebroventricular oleic acid decreased food intake 24 and 48 h later and prevented a fasting-induced elevation in hypothalamic NPY mRNA expression [31]. They have further demonstrated that the effects of oleic acid are nutritionally regulated. Three days of voluntary overfeeding, i.e. offering a palatable, energy-dense diet, blocked the effects of oleic acid on both food intake and NPY expression [32]. The authors interpreted

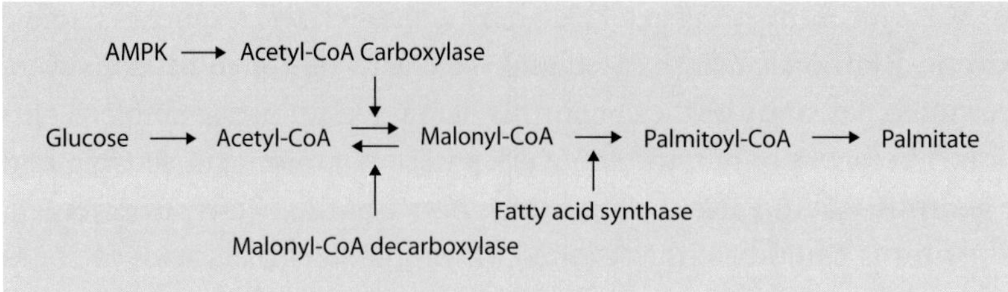

Fig. 1. Fatty acid synthetic pathway.

these data to suggest that the central availability of long-chain fatty acids provides a signal for metabolic excess.

A role for alterations in hypothalamic fatty acid synthesis in energy balance also derives from experiments examining the downstream actions of leptin [33]. Leptin has been shown to decrease pAMPK in the hypothalamus [34]. A downstream target of pAMPK is the rate-limiting enzyme for fatty acid synthesis, ACC (fig. 1). pAMPK increases pACC, the inactive form of ACC. Leptin-induced decreases in pAMPK resulted in decreases in pACC and increases ACC activity in both the hypothalamic paraventricular and ARC nuclei [34]. The increase in ACC activity led to elevated malonyl CoA in the ARC and to an increase in long-chain fatty acid CoA in the paraventricular nucleus.

Controversy exists over what aspect of fatty acid availability may inhibit eating. A number of proposals have been made. Rosetti and colleagues [31, 35] have proposed that a build-up of long-chain fatty acid CoA could serve as a signal of metabolic excess. In this view, either the exogenous administration of oleic acid or the actions of malonyl CoA as a CPT-1 inhibitor and the resulting increase in fatty acid CoA provide a signal of metabolic excess and lead to reductions in food intake. Consistent with this view are reports of eating inhibition by CPT-1 inhibitors [36] (although data to the contrary have also been reported [26]). Additional evidence for this view derives from the demonstration of increased food intake and body weight in mice overexpressing malonyl CoA decarboxylase (MCD; fig. 1) in the ARC [37]. MCD overexpression biases against malonyl CoA production, which may increase fatty acid oxidation; this decreased malonyl CoA signal and decreased inhibition of CPT-1 then results in increased food intake and increased body weight.

Malonyl CoA as a signal for inhibiting eating either alone or in conjunction with the brain form of CPT-1 (CPT-1c) has also been proposed [38–40]. Manipulations that increase malonyl CoA levels within the hypothalamus decrease food intake [23, 37, 41]. As noted above, lowering malonyl CoA levels by increasing the expression of MCD increases food intake and produces obesity [37]. CPT-1c is highly enriched at hypothalamic sites involved in the control of eating, and CPT-1c binds malonyl CoA but does not catalyze fatty acid transfer across the mitochondrial membrane [42]. Thus, malonyl CoA's interaction with CPT-1c does not affect β oxidation [43]. Wolfgang and colleagues have proposed that these characteristics suggest a regulatory function for CPT-1c, such that its interaction with malonyl CoA provide a signal for energy excess [42, 43]. However, knockout of CPT-1c results in a lean phenotype on chow but an obese phenotype on high fat diet, questioning how such a role may be mediated [42].

Hypothalamic Protein Sensing

Mammalian target of rapomycin (mTOR) is a serine-threonine kinase whose activity fluctuates in relation to nutrient status. For example, when intracellular energy stores are adequate, hypothalamic mTOR activity increases as AMPK activity is decreased [44]. mTOR appears to be involved in the anorectic actions of leptin and insulin [44, 45]. Consumption of a high-fat diet decreases mTOR signaling in the hypothalamus, which has been proposed to play a role in leptin and insulin insensitivity in animals maintained on a high fat diet [46]. The mTOR pathway, in addition to its responsiveness to adiposity signals, glucose and fatty acids, is sensitive to amino acids, especially branched-chain amino acids [47]. For example, leucine administered into the 3rd ventricle activates mTOR and reduces food intake [44].

High-protein diets have long been known to decrease total food intake and lead to weight loss [48]. Recent work has implicated AMPK and mTOR signaling in these effects [49]; i.e. feeding a high-amino acid or leucine-supplemented diet decreased overall food intake and body weight, reduced fat content and increased UCP-1 expression in brown adipose tissue. These alterations were accompanied by decreased hypothalamic AMPK activity and increased hypothalamic mTOR activity as well as decreased NPY and increased POMC expression within ARC. Furthermore, in another study, increasing amino acid or leucine concentrations in the brain decreased food intake and decreased ARC AgRP gene expression levels [50]. These actions appear to be mediated by increased activation of mTOR signaling as the addition of rapamycin prevented the effects in a cell culture preparation. Whether such alterations signal protein availability from meal to meal has yet to be demonstrated.

Conclusion

Overall, the data reviewed here demonstrate the capacity of hypothalamic neurons to monitor local changes in nutrient concentration and to alter their activity in response to these changes. Hypothalamic neurons that are responsive to leptin and insulin also often have the capacity of responding to specific nutrients. Although physiological alterations in glucose concentration are sensed, this does not appear to have direct effects on eating. Alterations in fatty acid metabolism are involved in mediating the eating- inhibitory actions of leptin and metabolic challenges. The mechanisms through which such alterations signal remain controversial. Recent work has suggested a role for mTOR signaling in the eating-inhibitory effects of high-protein diets and branched-chain amino acids. Such actions may also be involved in the signaling cascades initiated by glucose and fatty acids.

Acknowledgement

Aspects of this work were supported by DK19302.

References

1 Mayer J: Glucostatic mechanism of regulation of food intake. N Engl J Med 1953;249:13–16.
2 Oomura Y, Kimura K, Ooyama H, Maeo T, Iki M, Kuniyoshi N: Recipricol activites of the ventromedial and lateral hypothalamic areas in cats. Science 1964;143:484–485.
3 Fioramonti X, et al: Characterization of glucosensing neuron subpopulations in the arcuate nucleus: integration in neuropeptide Y and pro-opio-melanocortin networks? Diabetes 2007;56:1219–1227.
4 Levin BE, Dunn-Meynell AA, Routh VH: Brain glucosensing and the K(ATP) channel. Nat Neurosci 2001;4:459–460.
5 Levin B: Neuronal glucose sensing: still a physiological orphan. Cell Metab 2007;6:252–254.
6 Wang R, et al: The regulation of glucose-excited neurons in the hypothalamic arcuate nucleus by glucose and feeding-relevant peptides. Diabetes 2004;53: 1959–1965.
7 Spanswick D, et al: Leptin inhibits hypothalamic neurons by activation of ATP-sensitive potassium channels. Nature 1997;390:521–525.
8 Spanswick D, et al: Insulin activates ATP-sensitive K+ channels in hypothalamic neurons of lean, but not obese rats. Nat Neurosci 2000;3:757–758.
9 Stunkard AJ, Wolff HG: Studies on the physiology of hunger. I. The effect of intravenous glucose administration on gastric hunger contractions in man. J Clin Invest 1956;35:954–963.
10 Smith GP, Epstein AN: Increased feeding in response to decreased glucose utilization in the rat and monkey. Am J Physiol 1969;217:1083–1087.
11 Miselis RR, Epstein AN: Feeding induced by intracerebroventricular 2-deoxy-D-glucose in the rat. Am J Physiol 1975;229:1438–1447.
12 Smith G, Gibbs J, Stromayer AJ, Stokes PR: Threshold doses of 2-deoxy-D-glucose for hyperglycemia and feeding in rats and monkeys. American Journal of Physiology 1972;222:77–81.
13 Ritter S, Dinh TT, Li AJ: Hindbrain catecholamine neurons control multiple glucoregulatory responses. Physiol Behav 2006;89:490–500.
14 Campfield LA, Brandon P, Smith FJ: On-line continuous measurement of blood glucose and meal pattern in free-feeding rats: the role of glucose in meal initiation. Brain Res Bull 1985;14:605–616.
15 Campfield LA, Smith FJ: Transient declines in blood glucose signal meal initiation. Int J Obes 1990; 14(suppl 3):15–31, discussion 31–34.
16 Melanson KJ, et al: Blood glucose patterns and appetite in time-blinded humans: carbohydrate versus fat. Am J Physiol 1999;277:R337–R345.
17 Smith FJ, Campfield LA: Meal initiation occurs after experimental induction of transient declines in blood glucose. Am J Physiol 1993;265:R1423–R1429.
18 Dunn-Meynell AA, et al: Relationship among brain and blood glucose levels and spontaneous and glucoprivic feeding. J Neurosci 2009;29:7015–7022.
19 Parton LE, et al: Glucose sensing by POMC neurons regulates glucose homeostasis and is impaired in obesity. Nature 2007;449:228–232.
20 Oomura Y, et al: Effect of free fatty acid on the rat lateral hypothalamic neurons. Physiol Behav 1975; 14:483–486.
21 Wang R, et al: Effects of oleic acid on distinct populations of neurons in the hypothalamic arcuate nucleus are dependent on extracellular glucose levels. J Neurophysiol 2006;95:1491–1498.
22 Le Foll C, Irani BG, Magnan C, Dunn-Meynell AA, Levin BE: Characteristics and mechanisms of hypothalamic neuronal fatty acid sensing. Am J Physiol Regul Integr Comp Physiol 2009;297:R655–R664.
23 Loftus TM, et al: Reduced food intake and body weight in mice treated with fatty acid synthase inhibitors. Science, 2000;288:2379–2381.

24 Miller I, et al: Anorexigenic C75 alters c-Fos in mouse hypothalamic and hindbrain subnuclei. Neuroreport 2004;15:925–929.
25 Kim EK, et al: Expression of FAS within hypothalamic neurons: a model for decreased food intake after C75 treatment. Am J Physiol Endocrinol Metab 2002;283:E867–E879.
26 Aja S, et al: Pharmacological stimulation of brain carnitine palmitoyl-transferase-1 decreases food intake and body weight. Am J Physiol Regul Integr Comp Physiol 2008;294:R352–R361.
27 Yang N, et al: C75 [4-methylene-2-octyl-5-oxo-tetrahydro-furan-3-carboxylic acid] activates carnitine palmitoyltransferase-1 in isolated mitochondria and intact cells without displacement of bound malonyl CoA. J Pharmacol Exp Ther 2005;312:127–133.
28 Thupari JN, et al: C75 increases peripheral energy utilization and fatty acid oxidation in diet-induced obesity. Proc Natl Acad Sci USA 2002;99:9498–9502.
29 Mera P, et al: C75 is converted to C75-CoA in the hypothalamus, where it inhibits carnitine palmitoyltransferase 1 and decreases food intake and body weight. Biochem Pharmacol 2009;77:1084–1095.
30 Kim EK, et al: C75, a fatty acid synthase inhibitor, reduces food intake via hypothalamic AMP-activated protein kinase. J Biol Chem 2004;279: 19970–19976.
31 Obici S, et al: Central administration of oleic acid inhibits glucose production and food intake. Diabetes 2002;51:271–275.
32 Morgan K, Obici S, Rossetti L: Hypothalamic responses to long-chain fatty acids are nutritionally regulated. J Biol Chem 2004;279:31139–31148.
33 Gao S, et al: Leptin activates hypothalamic acetyl-CoA carboxylase to inhibit food intake. Proc Natl Acad Sci USA 2007;104:17358–17363.
34 Minokoshi Y, et al: AMP-kinase regulates food intake by responding to hormonal and nutrient signals in the hypothalamus. Nature 2004;428:569–574.
35 Lam TK, et al: Hypothalamic sensing of circulating fatty acids is required for glucose homeostasis. Nat Med 2005;11:320–327.
36 Obici S, et al: Inhibition of hypothalamic carnitine palmitoyltransferase-1 decreases food intake and glucose production. Nat Med 2003;9:756–761.
37 He W, et al: Molecular disruption of hypothalamic nutrient sensing induces obesity. Nat Neurosci 2006; 9:227–233.
38 Dai Y, et al: Localization and effect of ectopic expression of CPT1c in CNS feeding centers. Biochem Biophys Res Commun 2007;359:469–474.
39 Hu Z, et al: A role for hypothalamic malonyl-CoA in the control of food intake. J Biol Chem 2005; 280: 39681–39683.
40 Wolfgang MJ, Lane MD: Hypothalamic malonyl-coenzyme A and the control of energy balance. Mol Endocrinol 2008;22:2012–2020.
41 Lane MD, Cha SH: Effect of glucose and fructose on food intake via malonyl-CoA signaling in the brain. Biochem Biophys Res Commun 2009;382:1–5.
42 Wolfgang MJ, et al: Brain-specific carnitine palmitoyl-transferase-1c: role in CNS fatty acid metabolism, food intake, and body weight. J Neurochem 2008;105:1550–1559.
43 Wolfgang MJ, et al: The brain-specific carnitine palmitoyltransferase-1c regulates energy homeostasis. Proc Natl Acad Sci USA 2006;103:7282–7287.
44 Cota D, et al: Hypothalamic mTOR signaling regulates food intake. Science 2006;312:927–930.
45 Wullschleger S, Loewith R, Hall MN: TOR signaling in growth and metabolism. Cell 2006;124:471–484.
46 Cota D, et al: The role of hypothalamic mammalian target of rapamycin complex 1 signaling in diet-induced obesity. J Neurosci 2008;28:7202–7208.
47 Kimball SR, Jefferson LS: Signaling pathways and molecular mechanisms through which branched-chain amino acids mediate translational control of protein synthesis. J Nutr 2006;136(1 suppl):227S–231S.
48 Potier M, Darcel N, Tome D: Protein, amino acids and the control of food intake. Curr Opin Clin Nutr Metab Care 2009;12:54–58.
49 Ropelle ER, et al: A central role for neuronal AMP-activated protein kinase (AMPK) and mammalian target of rapamycin (mTOR) in high-protein diet-induced weight loss. Diabetes 2008;57:594–605.
50 Morrison CD, et al: Amino acids inhibit Agrp gene expression via an mTOR-dependent mechanism. Am J Physiol Endocrinol Metab 2007;293: E165–E171.

Timothy H. Moran, PhD
Department of Psychiatry and Behavioral Sciences, Johns Hopkins University School of Medicine, Ross 618
720 Rutland Ave.
Baltimore, MD 21205 (USA)
Tel. +1 410 955 2344, Fax +1 410 502 3769, E-Mail tmoran@jhmi.edu

Langhans W, Geary N (eds): Frontiers in Eating and Weight Regulation.
Forum Nutr. Basel, Karger, 2010, vol 63, pp 102–110

Blood-Brain Barrier as a Regulatory Interface

William A. Banks

GRECC, Veterans Affairs Medical Center, St. Louis and Saint Louis University School of Medicine, Division of Geriatrics, Department of Internal Medicine, St. Louis, Mo., USA

Abstract

The blood-brain barrier (BBB) is an important component of the communication network that connects the central nervous system and peripheral tissues in the control of feeding-related behaviors. Specifically, the BBB acts as an interface that restricts and regulates the exchange of substances between the CNS and blood. Many of the eating-related peptides and regulatory proteins produced by peripheral tissues and with receptors in the brain have been found to cross the BBB. The consequences of BBB permeability to these substances can be viewed from various perspectives. Here, we briefly consider five views relating the BBB and eating. A view of physiologic integration emphasizes the BBB as a conduit that controls a humoral-dependent signaling between the CNS and peripheral tissues. A view of regulation emphasizes that the transporters for many of the eating-related hormones are themselves regulated by physiologic events. This means that blood-to-brain signaling across the BBB is state-dependent and adaptable to the needs of the organism. A view of pathologic dysfunction shows how dysregulation of BBB transporters can result in disease. Resistance to leptin caused by its decreased transport across the BBB in obesity is an example. An evolutionary view emphasizes how the role of the BBB in eating may have evolved and how adaptations to one set of eating conditions can result in maladaptations under other conditions. Finally, the implications of these views for drug development targeted at obesity or anorexia is explored. Overall, these views show the BBB is an integral part of the physiology of eating.

The simple act of eating belies a highly complex, highly integrated network of signals and events that occur on a number of fronts, as introduced in this volume [1]. From social considerations to coordination of gastrointestinal musculature, from psychological associations with eating to regulation of parietal cell secretions, eating is preceded and followed by events that must be coordinated one with the other. Although the area of eating cuts across many fields of learning, coordination involves only two basic mechanisms: neural and endocrine. These mechanisms act independently and together to coordinate actions occurring within and between the central nervous system (CNS) and peripheral tissues.

Humoral communication between the CNS and peripheral tissues is, however, complicated by the blood-brain barrier (BBB), which limits and regulates the exchange of substances between the fluids of the CNS (brain interstitial fluids/cerebrospinal fluid) and the blood. This brief review will focus on the roles of the BBB in regulation of eating by considering physiologic, regulatory, pathologic, evolutionary, and therapeutic views.

A View of Physiologic Integration

The most striking aspect of BBB function in eating is its ability to transport the relatively large peptides and regulatory proteins from blood to brain. The BBB was originally defined on the basis of its ability to exclude serum proteins such as albumin from leaking across the capillary bed of the brain into brain interstitial fluid [2]. This was later extended to molecules as small as the element lanthanum (138.91 Da), which was used in electron microscopy studies to visualize the tight junctions between endothelial cells. That large molecules such as leptin (16,000 Da) can cross the BBB seems at first against the dogma of the BBB excluding proteinaceous material. However, the presence of specific transporters at the BBB allows for the blood-to-brain transport of eating-related peptides and large regulatory proteins.

The first eating-related hormone whose BBB transport was considered was insulin, with original studies published in the 1950s. Those studies concluded that little or no insulin crossed the BBB [3, 4]. The question was later revived by Woods and Porte [5], whose work convinced many that insulin crossed the BBB by way of a saturable transport system. Melanocyte-simulating hormone (MSH) was proposed to cross the BBB; however, it was studies of its cognitive effects that led to this proposal and not its effects on eating [6]. MSH was shown to cross the BBB by a nonsaturable mechanism in the 1980s [7]. Starting in 1989, a number of cytokines (e.g. interleukins, tumor necrosis factor-α) were shown to cross the BBB by way of saturable transporters [8], but during this early period the emphasis on study was on immune function, not eating. The idea that peptides and regulatory proteins are transported across the BBB was highly controversial throughout the 1980s and well into the 1990s.

Interest in the BBB and eating-related hormones increased dramatically with the discovery of leptin. Despite its size of about 16 kDa, it was assumed from nearly the beginning that leptin would need to cross the BBB to exert its effects on eating in the brain [9]. For a molecule this large, passage would likely require a saturable transporter, and this was shown formally in 1996 [10].

Since then, a long list of eating-related hormones has since been shown to cross the BBB by either saturable or nonsaturable mechanisms. This includes ghrelin, amylin, secretin, pancreatic polypeptide, galanin-like peptide, and orexin A [11]. To date, no eating-related peptide or protein has been shown to be transported in the brain-to-blood direction by a saturable system. In contrast, the enzymatic activity of the BBB prevents monoamines related to eating from entering the CNS from blood and steroid hormones cross the BBB by nonsaturable mechanisms [12, 13].

A View of Regulation

The transport of eating-related peptides and regulatory proteins across the BBB is not static but is affected by physiologic effects. The transport into brain of regulatory proteins shows diurnal rhythms [14, 15]. The transport of leptin, insulin, and ghrelin are affected by obesity, glucose, fasting/starvation, serum triglycerides, and adrenergics [16–21]. The transporter for insulin is turned off during hibernation in marmots [22]. These findings illustrate the point that the transporters are capable of responding to physiologic needs.

Eating-related hormones also regulate one another. Insulin enhances leptin transport across the BBB, and leptin induces a saturable transporter for urocortin [15, 23]. The role that these interactions among eating-related hormones play is unclear, but it opens a new area of investigation and raises fresh questions about the physiologic interplay among these hormones.

The BBB possesses aspects other than the yin and yang of barrier and permeability. At least three other functions of the BBB are important to understanding it regulatory aspects: enzymatic activity, possession of receptors that modulate its cellular responses, and secretions. The study of secretions by the BBB is a relatively new field with fewer than 40 publications. Brain endothelial cells and ependymal cells of the choroid plexus can secrete prostaglandins, nitric oxide, and cytokines [24–26]. Such secretions can be constitutive or induced. Interleukin-6 (IL-6) is the best studied with regard to its secretion by the BBB. It is constitutively expressed but its secretion is greatly increased with exposure to lipopolysaccharide (LPS) [27]. Secretions can be from either the luminal (blood) or abluminal (brain) side of the brain and stimulation on one side can induce release from the other [28]. Brain endothelial cells possess receptors to adiponectin and recent work indicates that adiponectin can inhibit IL-6 secretion from these cells [29]. IL-6 is inversely related to adiposity and seems to exert effects on body fat mass [30]. Regulation of IL-6 release from brain endothelial cells into the brain could be a mechanism by which circulating adiponectin affects adiposity without adiponectin itself, which has lower CSF/serum ratios than even albumin [29, 31], having to cross the BBB.

A View of Pathologic Dysfunction

The above views portray the BBB not only as a barrier restricting movement of proteins from blood to brain, but also as a regulatory interface between the circulation and the CNS. This latter aspect emphasizes the ability of the BBB to regulate the exchange of informational molecules between the CNS and peripheral tissues (via the blood) and so to act as the crossroads in a humoral-based communication between the brain and body. Failure of such a communication system would likely lead to disease. Such a failure occurs in obesity.

Resistance to the actions of leptin is associated with obesity in humans [32, 33]. Classic hormone resistance was first described for pseudohypoparathyroidism [34]. Under normal conditions, calcium levels in serum are normally tightly regulated by a negative-feedback loop between secretion of parathyroid hormone (PTH) and calcium. In pseudohypoparathyroidism, a dysfunction of the PTH receptor results in hypocalcemia. If the receptor resistance is partial, then the parathyroid tissue simply secretes more PTH (which may involve hypertrophy of the parathyroid glands) until calcium levels are brought into the normal range. In obesity, the effector molecule is leptin, not PTH, and the regulated endpoints are eating and thermogenesis, not calcium. Overcoming resistance means secreting more leptin which is achieved by increasing fat mass. The leptin-eating/thermogenesis feedback loop, however, is more complicated than the parathyroid-calcium loop in that there are two points in addition to the receptor/postreceptor level at which resistance can arise: the saturable transport at the BBB and downstream neuronal circuitries (that is, effects on secretion and function of anorexins and orexins within the brain). Leptin resistance could, in theory, be produced at any of these three points: BBB transport; leptin receptor/postreceptor function; downstream neuronal circuitries.

Evidence shows that impaired BBB transport of leptin occurs in obesity [35, 36] (other mechanisms are discussed by Munzberg-Gruenig [37] in this volume). This is, in part, because the levels of serum leptin in obesity partially saturate the BBB transporter. This means that as serum leptin levels increase, a lesser percent of serum leptin is transported across the BBB. Additionally, regulation of the leptin transporter is likely involved in inducing resistance at the BBB [19]. Triglycerides, for example, inhibit the transport of leptin across the BBB [16]. Triglycerides are elevated in obesity and so may contribute to the leptin resistance [38].

Other circulating factors also likely affect the transporters for leptin and other eating-related hormones. Ghrelin transport, for example, appears to be influenced by serum factors in the feeding and fasting state [20]. Triglycerides promote the transport of insulin across the BBB [21]. Thus, triglycerides could tend to promote feeding both by inhibiting the BBB transport of the anorexic leptin and stimulating the BBB transport of the orexigen ghrelin.

An Evolutionary View

Evolutionary pressures adapt the characteristics of a breeding population to maximize reproductive success. When the evolutionary pressures shift, adaptations may become maladaptations and can manifest as disease. Leptin seems ill-fitted as a regulator of adiposity given that resistance readily develops at the BBB and the receptor and that obesity-related triglycerides promote orexigenic rather than anorexic signals from blood to brain. These and other findings suggest that leptin is not an adiposity signal,, that is, its primary purpose is not to regulate adiposity.

Analysis of the Paleolithic diet suggests that simple carbohydrates and fat were always in short supply. Behaviors and physiology that would amass calories and fat would increase reproductive success. Consistent with this, baboons living in the wild have body fat masses that are 2–3% of total body weight and leptin levels that are 1–2 ng/ml [39, 40]. In comparison, body fat constitutes 10–20% of a person at ideal body weight, and leptin levels for such a person are about 10 ng/ml. Therefore, the leptin signaling system likely evolved in ancestors with low fat reserves and low serum leptin levels. The evolutionary need, then, would have been for a signal to brain of the level of those fat reserves. When those reserves were low, the brain could conserve energy by limiting calorically demanding expenditures such as supporting the immune system and reproduction. When the brain sensed adequate caloric reserves, behaviors other than those related to eating could be indulged and resources invested in the immune system and reproduction. This view can reconcile some otherwise mysterious aspects of leptin physiology. It provides an explanation of why the most efficient transport of leptin across the BBB occurs at low blood levels of leptin and why those low blood levels of leptin have stimulatory or permissive effects on functions seemingly unrelated to eating, such as the immune system and reproduction.

Why should hypertriglyceridemia favor orexigenic rather than anorexigenic signaling across the BBB? Elevations in serum triglycerides occur not only in obesity, but also in starvation. Starvation has likely been much more common than obesity during evolutionary history and the consequences of starvation on reproductive success are much more profound than those of obesity. Indeed, seasonal nutritional stress is a regular feature of life among extant hunter-gathers and is a threat to the survival of those with inadequate reserves of fat [41]. Hypertriglyceridemia may have evolved, therefore, as an indirect signal to the brain of starvation by inhibiting the transport of the anorexin leptin and promoting the BBB transport of the orexin ghrelin.

A Drug Delivery View

The complexity of the physiological control of eating presents many targets for therapeutic intervention. This complexity also produces a resilience that often defeats a single intervention. Diet and exercise remain the cornerstones of successful, sustained weight loss, but the bias of the eating system towards preservation of caloric reserves means that patients must be dedicated and resourceful to maintain initial weight loss. The complexities and difficulties have led many to believe that targeting the hormones, peptides, and neurotransmitters that control body weight make drug development in this area near impossible. But an examination of the facts suggests that effective drug development is possible. First, many drugs can produce dramatic, successful and sustained weight loss, such as thyroxine, concentrations of the herbal ma huang (containing the ephedrine as an active ingredient), and fen-phen (fenfluramine and phentermine). Unfortunately, these drugs have serious side effects, especially on the cardiovascular system, and their abuse is associated with increased mortality. Nevertheless, their ability to induce weight loss can be taken as proof of principle. The challenge, then, is to produce drugs that target feeding without excess toxicity. Second, the amount of weight lost does not need to be dramatic, as loss of 5–10% of body weight is associated with improved cardiovascular and metabolic outcomes. Third, effects on food intake or metabolic rate are adjustments; a deficit or excess of 10 calories a day results in a weight change of a pound a year. Fourth, examples of weight loss (and weight gain) that are mediated by alterations in hormones in brain or blood are provided by some diseases.

Anorexia is a medical problem that in some populations is more threatening than obesity. Anorexia of aging, anorexia nervosa, and anorexia from various cancers are examples [42]. Whereas leptin resistances have so far limited the use of leptin in obesity, leptin deficiencies are associated with profound increases in adiposity. It may be that developing drugs that treat anorexia may prove easier than developing drugs that treat obesity. Recently, a pegylated leptin antagonist was found to increase body weight by blocking the transport of endogenous leptin across the BBB [43]. Likewise, targeting the melanocortin receptor MCR4, developing ghrelin antagonists, or combining therapies are possible strategies.

Often drug development targeted to one disease results in drugs for another. Many, if not most, of the peptides and regulatory proteins involved in feeding also have effects on cognition. These include melanocortin, leptin, insulin, and ghrelin [44–47]. The reasons for this close match are unclear. It may be that the complexity of eating served as the substrate from which a more complex form of cognition arose. Solving eating-related problems (cost/benefit ratio analysis of the likelihood that calories gathered will exceed calories expended in searching for food; risk/benefit ratio analysis of dangers of searching for food vs likelihood of finding food; complex social rules for hunting and sharing food) was likely an early application of enhanced cognition. Whatever the causal link, eating-related hormones have profound effects on brain development and cognition in the adult. Many of the feeding hormones show promise as therapeutic agents in the treatment of dementia [48, 49].

Summary: An Integrated View

An ultimate understanding of the interactions between the peripheral tissues and CNS in eating must integrate the BBB. That integration is multifaceted and includes physiologic, regulatory, pathologic, evolutionary, and therapeutic views. The BBB acts as an interface between the CNS and peripheral tissues, controlling the exchange of peptides and regulatory proteins. The transporters for eating-related hormones are, in turn, themselves regulated. Failure of BBB functions can lead to diseases as exemplified by obesity and impaired leptin transport. Functional aspects of the BBB, as exemplified by the leptin transporter, are consistent with the idea that eating-related systems evolved at body fat levels much lower than those considered normal in Western societies. This area also suggests that starvation was predominant over caloric excess as an evolutionary pressure in shaping the BBB transport of eating-related hormones. In the treatment of eating disorders, the BBB is a therapeutic target in its own right as well as the conduit that must deliver any therapeutics to the CNS. From these views and others, the BBB is an integral part in the feeding networks that connect the central nervous system and peripheral tissues.

References

1 Langhans W, Geary N: Overview of the physiological control of eating; in Langhans W, Geary N (eds): Frontiers in Eating and Weight Regulation. Forum Nutr. Basel, Karger, 2010, vol 63, pp 9–53.

2 Davson H: The blood-brain barrier; in: Physiology of the Cerebrospinal Fluid. London, Churchill, 1967, pp 82–103.

3 Elgee NJ, Williams RH, Lee ND: Distribution and degradation studies with insulin-I[131]. J Clin Invest 1954:33:1252–1260.

4 Haugaard N, Vaughan M, Haugaard ES, Stadie WC: Studies of radioactive injected labeled insulin. J Biol Chem 1954:208:549–563.

5 Woods SC, Porte D Jr.: Relationship between plasma and cerebrospinal fluid insulin levels of dogs. Am J Physiol 1977;233:E331–E334.

6 Kastin AJ, Coy DH, Schally AV, Miller LH: Peripheral administration of hypothalamic peptides results in CNS changes. Pharmacol Res Comm 1978;10:293–312.

7 Wilson JF, Anderson S, Snook G, Llewellyn KD: Quantification of the permeability of the blood-CSF barrier to α-MSH in the rat. Peptides 1984;5:681–685.

8 Banks WA, Kastin AJ: The interleukins-1α, -1β, and -2 do not disrupt the murine blood-brain barrier. Int J Immunopharmacol 1992;14:629–636.

9 Campfield LA, Smith FJ, Guisez Y, Devos R, Burn P: Recombinant mouse OB protein: evidence for a peripheral signal linking adiposity and central neural networks. Science 1995;269:546–549.

10 Banks WA, Kastin AJ, Huang W, Jaspan JB, Maness LM: Leptin enters the brain by a saturable system independent of insulin. Peptides 1996;17:305–311.

11 Kastin A J, Pan W. Blood-brain barrier and feeding: regulatory roles of saturable transport systems for ingestive peptides. Curr Pharmac Design 2008;14: 1615–1619.

12 Hardebo JE, Owman C: Enzymatic barrier mechanisms for neurotransmitter monoamines and their precursors at the blood-brain barrier; in Johansson BB, Owman C, Widner H (eds): Pathophysiology of the Blood-Brain Barrier. Amsterdam, Elsevier, 1990, pp 41–55.

13 Pardridge WM, Mietus LJ: Transport of steroid hormones through the rat blood-brain barrier: primary role of albumin-bound protein. J Clin Invest 1979; 64:145–154.

14 Banks WA, Kastin AJ, Ehrensing CA: Diurnal uptake of circulating interleukin-1α by brain, spinal cord, testis and muscle. NIM 1998;5:36–41.

15 Pan W, Kastin AJ: Diurnal variation of leptin entry form blood to brain involving partial saturation of the transport system. Life Sci 2001;68:2705–2414.

16 Banks WA, Coon AB, Robinson SM, Moinuddin A, Shultz JM, Nakaoke R, Morley JE: Triglycerides induce leptin resistance at the blood-brain barrier. Diabetes 2004;53:1253–1260.

17 Banks WA: Enhanced leptin transport across the blood-brain barrier by α1-adrenergic agents. Brain Res 2001;899:209–217.

18 Kastin AJ, Akerstrom V: Glucose and insulin increase the transport of leptin through the blood-brain barrier in normal mice but not in streptozotocin-diabetic mice. Neuroendocrinology 2001;73: 237–242.

19 Banks WA, DiPalma CR, Farrell CL: Impaired transport of leptin across the blood-brain barrier in obesity. Peptides 1999;20:1341–1345.

20 Banks WA, Burney BO, Robinson SM: Effects of triglycerides, obesity, and starvation on ghrelin transport across the blood-brain barrier. Peptides 2008;29:2061–2065.
21 Urayama A, Banks WA: Starvation and triglycerides reverse the obesity-induced impairment of insulin transport at the blood-brain barrier. Endocrinology 2008;149:3592–3597.
22 Florant GL, Richardson RD, Mahan S, Singer L, Woods SC: Seasonal changes in CSF insulin levels in marmots: insulin may not be a satiety signal for fasting in winter. Am J Physiol 1991;260:R712–R716.
23 Kastin AJ, Akerstrom V, Pan W: Activation of urocortin transport into brain by leptin. Peptides 2000; 21:1811–1817.
24 Fabry Z, Fitzsimmons KM, Herlein JA, Moninger TO, Dobbs MB, Hart MN: Production of the cytokines interleukin 1 and 6 by murine brain microvessel endothelium and smooth muscle pericytes. J Neuroimmunol 1993;47:23–34.
25 Mandi Y, Ocsovszki I, Szabo D, Nagy Z, Nelson J, Molnar J: Nitric oxide production and MDR expression by human brain endothelial cells. Anticancer Res 1998;18:3049–3052.
26 MCGuire TR, Trickler WJ, Hock L, Vrana A, Hoie EB, Miller DW: Release of prostaglandin E-2 in bovine brain endothelial cells after exposure to three unique forms of the antifungal drug amphotericin-B: role of COX-2 in amphotericin-B induced fever. Life Sci 2003;72:2581–2590.
27 Reyes TM, Fabry Z, Coe CL: Brain endothelial cell production of a neuroprotective cytokine, interleukin-6, in response to noxious stimuli. Brain Res 1999;851:215–220.
28 Verma S, Nakaoke R, Dohgu S, Banks WA: Release of cytokines by brain endothelial cells: a polarized response to lipopolysaccharide. Brain Behav Immun 2006;20:449–455.
29 Spranger J, Verma S, Gohring I, Bobbert T, Seifert J, Sindler AL, Pfeiffer A, Hileman SM, Tschop M, Banks WA: Adiponectin does not cross the blood-brain barrier, but modifies cytokine expression of brain endothelial cells. Diabetes 2006;55:141–147.
30 Bruun JM, Lihn AS, Verdich C, Pedersen SB, Toubro S, Astrup A, Richelsen B: Regulation of adiponectin by adipose tissue-derived cytokines: in vivo and in vitro investigations in humans. Am J Physiol Endocrinol Metab 2003;285:E527–E533.
31 Kusminski CM, McTernan PG, Schraw T, Kos K, O'Hare JP, Ahima R, Kumar S, Scherer PE: Adiponectin complexes in human cerebrospinal fluid: distinct complex distribution from serum. Diabetologia 2007;50:634–642.
32 Caro JF, Kolaczynski JW, Nyce MR, Ohannesian JP, Opentanova I, Goldman WH, Lynn RB, Zhang PL, Sinha MD, Considine RV: Decreased cerebrospinal-fluid/serum leptin ratio in obesity: a possible mechanism for leptin resistance. Lancet 1996;348: 159–161.
33 Schwartz MW, Peskind E, Raskind M, Boyko EJ, Porte D Jr: Cerebrospinal fluid leptin levels: relationship to plasma levels and adiposity in humans. Nat Med 1996;2:589–593.
34 Verhoeven GFM, Wilson JD: The syndromes of primary hormone resistance. N Engl J Med 1979;28: 253–289.
35 van Heek M, Compton DS, France CF, Tedesco RP, Fawzi AB, Graziano MP, Sybertz EJ, Strader CD, Davis J: Diet-induced obese mice develop peripheral, but not central, resistance to leptin. J Clin Invest 1997;99:385–390.
36 Halaas JL, Boozer C, Blair-West J, Fidahusein N, Denton DA, Friedman JM: Physiological response to long-term peripheral and central leptin infusion in lean and obese mice. Proc Natl Acad Sci USA 1997;94:8878–8883.
37 Münzberg H: Leptin-signaling pathways and leptin resistance; in Langhans W, Geary N (eds): Frontiers in Eating and Weight Regulation. Forum Nutr. Basel, Karger, 2010, vol 63, pp 123–132.
38 Banks WA: The blood-brain barrier as a cause of obesity. Curr Pharmac Design 2008;14:1606–1614.
39 Altmann J, Schoeller D, Altmann SA, Muruthi P, Sapolsky RM: Body size and fatness of free-living baboons reflect food availability and activity levels. Am J Primatol 1993;30:149–161.
40 Banks WA, Phillips-Conroy JE, Jolly CJ, Morley JE: Serum leptin levels in wild and captive populations of baboons (*Papio*): implications for the ancestral role of leptin. J Clin Endocrinol Metab 2001;86:4315–4320.
41 Eaton SB, Konner M: Paleolithic nutrition: a consideration of its nature and current implications. N Engl J Med 1985;312:283–289.
42 Morley JE: Anorexia of aging: physiologic and pathologic. Am J Clin Nutr 1997;66:760–773.
43 Elinav E, Niv-Spector L, Katz M, Price TO, Ali M, Yacobovitz M, Solomon G, Reicher S, Lynch JL, Halpern Z, Banks WA, Gertler A: Pegylated leptin antagonist is a potent orexigenic agent: preparation and mechanism of activity. Endocrinology 2009;150: 3083–3091.
44 Long JB, Rigamonti DD, Dosaka K, Kraimer JM, Martinez-Arizala A: Somatostatin causes vasoconstriction, reduces blood flow and increases vascular permeability in the rat central nervous system. J Pharmacol Exp Ther 1992;260:1425–1432.

45 Diano S, Farr SA, Benoit SE, McNay EC, da Silva I, Horvath B, Gaskin FS, Nonaka N, Jaeger LB, Banks WA, Morley JE, Pinto S, Sherwin RS, Xu L, Yamada KA, Sleeman MW, Tschop MH, Horvath TL: Ghrelin controls hippocampal spine synapse density and memory performance. Nat Neurosci 2006;9: 381–388.
46 Farr SA, Banks WA, Morley JE: Effects of leptin on memory processing. Peptides 2006;27:1420–1425.
47 Zhao WQ, Chen H, Quon MJ, Alkon DL: Insulin and the insulin receptor in expreimental models of learning and memory. Eur J Pharmacol 2004;490:71–81.
48 Reger MA, Watson GS, Green PS, Baker LD, Cholerton B, Fishel MA, Plymate SR, Cherrier MM, Schellenberg GD, Frey WH II, Craft S: Intranasal insulin administration dose-dependently modulates verbal memory and plasma amyloid-beta in memory-impaired adults. J Alzheimer Dis 2008;13:323–331.
49 Kern W, Born J, Schreiber H, Fehm HL: Central nervous system effects of intranasally administered insulin during euglycemia in men. Diabetes 1999;48: 557–563.

William A. Banks
915 N. Grand Blvd
St. Louis, MO 63106 (USA)
Tel. +1 314 289 7084, Fax +1 314 289 6374, E-Mail bankswa@slu.edu

Langhans W, Geary N (eds): Frontiers in Eating and Weight Regulation.
Forum Nutr. Basel, Karger, 2010, vol 63, pp 111–122

Do Leptin and Insulin Signal Adiposity?

Jacquelien J.G. Hillebrand · Nori Geary

Physiology and Behaviour Group, Institute of Food, Nutrition and Health, ETH Zürich, Schwerzenbach, Switzerland

Abstract

The physiological regulation of adiposity is supposed to depend on endocrine 'adiposity signals' that inform the brain about the mass of the adipose tissue. Basal levels of insulin and leptin are widely accepted to be adiposity signals, and amylin, ghrelin and peptide YY have been hypothesized to be. Support for these ideas comes from associations between basal hormone levels and levels of adiposity, from demonstrations of receptors for these hormones in neural circuits supposed to regulate energy homeostasis, from neuropharmacological manipulations of the hormones' actions on eating and energy expenditure, and from the effects on energy balance in animals or people bearing mutations in these endocrine signaling pathways. This chapter focuses on only the first of these four types of evidence and only on insulin and leptin. We ask whether circulating levels of either hormone indeed encodes the necessary information to act as an adiposity signal. In considering this question, we emphasize the distinction between regulation of AT mass in steady versus dynamic states. We argue that the best experimental designs for identifying potentially effective adiposity signals involve situations in which the level of adiposity is changing as the organism responds to imposed perturbations. Traditionally, this is the type of design that most convincingly supports the idea that adiposity is actively regulated. Unfortunately, there are few of such studies for any of the hypothesized endocrine adiposity signals, and the evidence that is available does not strongly support the hypotheses. Therefore, we conclude that the question of how adiposity is signaled to the brain remains an open frontier in the physiology of energy homeostasis.

Introduction

Energy Homeostasis

Several lines of evidence indicate that active regulation of adiposity contributes to energy homeostasis [1–9]. The aspect of adiposity that is regulated appears to be adipose tissue (AT) mass. Because body weight and body mass index (BMI) are often poor predictors of AT mass in rodents and humans [10–12], we emphasize studies in which AT mass is directly estimated.

In addition to the regulation of AT mass, the regulation of immediately available metabolic fuel is also an important component of energy homeostasis [5]. This may be described as the regulation of energy flow. It is important to realize that only chronic, consistent changes in energy flow affect AT mass and that in the short term, the two can vary independently.

Active Regulation of AT

A frequently cited piece of evidence for the existence of regulation of AT mass is the accuracy with which energy intake and expenditure must be matched to avoid the large increases in AT mass that would theoretically ensue if energy intake consistently exceeded energy expenditure (EE). As has been pointed out years ago [13], however, constancy per se does not reveal whether regulation results from an equilibrium among counterbalancing factors or from an active mechanism. Two further lines of evidence, more directly indicate that AT mass is regulated by an active system. The first is that robust compensatory responses follow experimentally induced underweight or overweight [6, 14]. That is, experimentally induced increases in body weight, and therefore AT mass, almost invariably lead to coordinate decreases in eating and increases in EE, and experimentally induced decreases in body weight to the opposite responses. Furthermore, the magnitudes of the compensatory responses often parallel the degree of weight perturbation. These kinds of observations convincingly support the idea of regulation of AT mass by an active, negative-feedback mechanism.

The second line of evidence indicating that AT mass is actively regulated is that for over 60 years there have been plausible mechanistic bases for its regulation, especially neurological analyses of hypothalamic control of energy balance [1–9]. The initial neurological evidence for active regulation of AT came from analyses of ventromedial hypothalamic and lateral hypothalamic lesion syndromes. More recently, analyses of single gene mutation obesity syndromes have revealed an extensive neuronal network subserving the regulation of AT [5, 15, 16].

Adiposity Signals

What feedback signal or signals link the AT to the brain? These feedback signals have long been thought to be humoral factors [4, 17, 18]. Although metabolites released by the AT, such as glycerol [19], were formerly considered viable candidate signals, these now seem rather weak ones [6]. Instead, these days basal or average levels of certain hormones, especially insulin and leptin, are thought to signal adiposity. These hormones are referred to as adiposity signals. The place of these negative-feedback signals in the regulation of AT mass is shown schematically in figure 1a. Given the

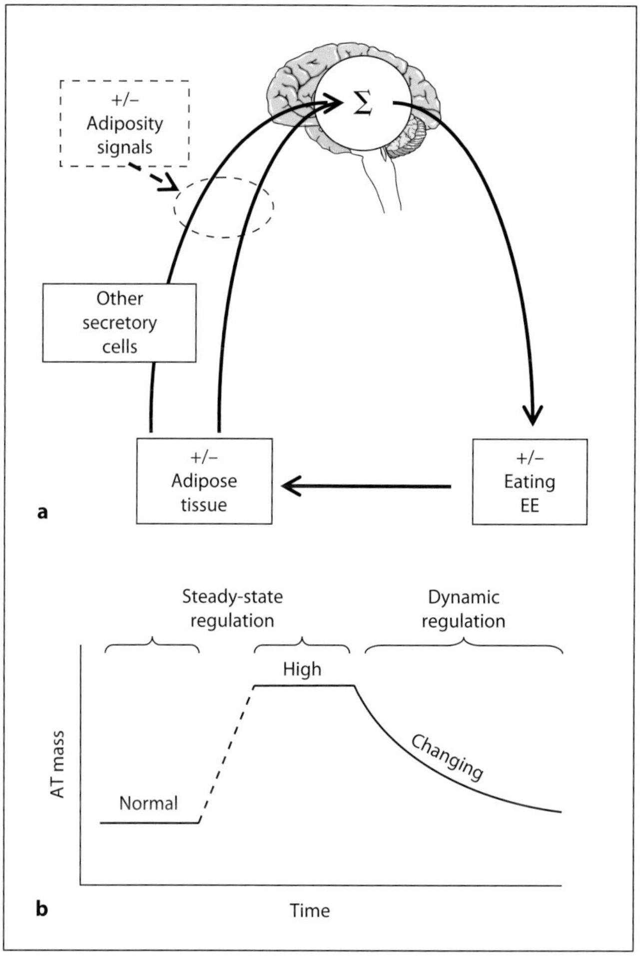

Fig. 1. **a** Schematic of the hypothetical negative-feedback homeostatic regulation of adipose tissue (AT) mass. Adipocytes secrete leptin and other secretory cells secrete insulin, amylin, peptide YY, ghrelin or other endocrine adiposity signals that encode AT mass. These are sensed and integrated (Σ) with the result that eating and energy expenditure (EE) are adjusted so as to maintain AT mass at the appropriate level (sometimes called the set point). For further discussion, see Langhans and Geary [5]. **b** Schematic illustration of the distinction between steady-state and dynamic regulation of AT mass. Steady-state regulation maintains AT mass at constant levels; depending on other variables, e.g. food restriction or the availability of highly palatable, energy-dense diets, maintenance can occur at low, normal, or high levels. Dynamic regulation refers to the robust eating and EE responses that return AT mass from the perturbed levels to normal level.

degree of confidence often expressed about the physiological roles of insulin and leptin as adiposity signals [1, 7, 8, 20], it is not an exaggeration to say that this view is a current physiological dogma. The theme of our review is to re-consider one aspect of the support for this dogma, namely, whether insulin and leptin reflect AT mass sufficiently accurately to serve as adiposity signals. Amylin [21], ghrelin [22, 23], and peptide YY (PYY) [24] are also candidate endocrine adiposity signals, but space limitations preclude us reviewing them here.

The Problem

Endocrine Coding

A variety of criteria are prerequisite for establishing physiological roles of hormones [5, 23, 25]. Perhaps the most fundamental of these is that hormone concentrations reaching the critical receptors encode the information signaled. This may be called the endocrine coding criterion. As reviewed below, in the case of adiposity signals, this criterion is usually taken to mean that a hormone is a plausible adiposity signal if basal plasma levels of the hormone are correlated to AT mass [7, 8, 23]. ('Basal' usually means after overnight fasting in humans and after overnight or shorter fasts in rodents.) This definition, however, appears incomplete, for at least two reasons. First, because the critical receptors for adiposity signals are in the brain, changes in blood-brain-barrier function may significantly alter the relation between plasma levels of a hormone and the levels that reach the critical receptors. This issue is discussed by Banks [26] in this volume. Second, the coding criterion has been assumed to be fulfilled by correlations between basal plasma levels and AT mass in situations in which AT mass is constant. We argue here that this is an incomplete test of the criterion.

Tests of the Endocrine Coding Criterion for Adiposity Signals

The majority of support for endocrine adiposity signals comes from cross-sectional studies during steady-state conditions. For example, normal-weight and obese humans are compared, or the effects of manipulations of AT mass produced by feeding palatable, energy-dense diets or by gene mutations, etc., are compared. Because subjects are in steady-state situations, AT mass, levels of feedback signals, and the intensity of regulatory responses are all relatively constant. In our view, these designs produce only weak evidence that the hypothesized adiposity signal functions as a feedback signal that initiates active regulation of AT mass and, therefore, do not alone fulfill the endocrine coding criterion. We believe that much stronger evidence would come from studies of changes in hypothesized adiposity signals in situations in which the organism is dynamically regulating AT mass, i.e. when it is responding to perturbations of AT mass by over- or under-eating or by increasing or decreasing EE (fig. 1b). As reviewed above, it is such dynamic situations that produce the most compelling evidence for regulation of AT mass in the first place.

It is important to distinguish tests involving AT-regulatory responses from shorter-term tests of energy flow. In many studies, energy intake is manipulated acutely, e.g.

by fasts, overfeeding, etc., for 1–3 days. Such manipulations can lead to significant changes in body weight, but these weight changes usually reflect hydration, gut contents, etc., with little or no change in AT mass. Nevertheless, marked changes in leptin, insulin, and other hypothesized adiposity signals often occur, both in animals [27, 28] and humans [29, 30]. Our interpretation of such data is that they are not relevant to the evaluation of adiposity signals because they involve dynamic changes in energy flow in the absence of changes in AT mass. Therefore, we do not consider them here.

Insulin as an Adiposity Signal

The plausibility of insulin as an adiposity signal was established in the 1960–70s by Porte and his colleagues, who demonstrated that basal plasma insulin levels are positively correlated with percent ideal body weight in humans [31] and with body weight in rats [32] (fig. 2a, b). These data led Woods et al. [33] to the hypothesis that basal insulin is an adiposity signal; the current version of this hypothesis now enjoys strong support from each of the types of evidence described in Langhans and Geary [5] and Woods et al. [20].

In the 1980s, Polonsky and colleagues added important details to the relation between insulin and AT mass. Using plasma C-peptide levels to evaluate basal and prandial insulin secretion rates in normal-weight and obese humans, they verified that the earlier data reflected changes in basal insulin secretion and also demonstrated that, whereas insulin secretion returns to the basal level in normal-weight subjects, it remains elevated before lunch and dinner in obese subjects [34, 35].

Myriad cross-sectional and longitudinal studies in humans and rats in steady states indicate that basal plasma insulin is increased or decreased in the appropriate direction in humans or rats who are spontaneously heavier or lighter or after weight manipulations, which also supports the hypothesis that basal insulin encodes AT mass [8, 9, 20, 31, 32]. For example, MacLean et al. [36] reported positive correlations between basal plasma insulin levels and retroperitoneal and epididymal AT mass in rats whose body weights were manipulated by repeated periods of palatable-food access or food restriction. There are also exceptions. For example, Korner et al. [37] reported that basal insulin levels in Roux-en-Y gastric bypass patients at ≥1 year of follow-up were only ~40% of those in BMI-matched controls. This may make sense in view of the patients' improved insulin function, but nevertheless dissociates insulin levels from AT.

Because of the lack of studies on levels of adiposity signals while rats are dynamically compensating for forced underweight or overweight, we recently began such an investigation. Rats were restrictively fed until they had ~23 g less AT versus normal-weight controls or intragastrically overfed until they had ~103 g more AT, measured by computed tomographic scanning (fig. 3). When these interventions were stopped, the rats displayed prolonged, dynamic regulatory responses that reduced the AT

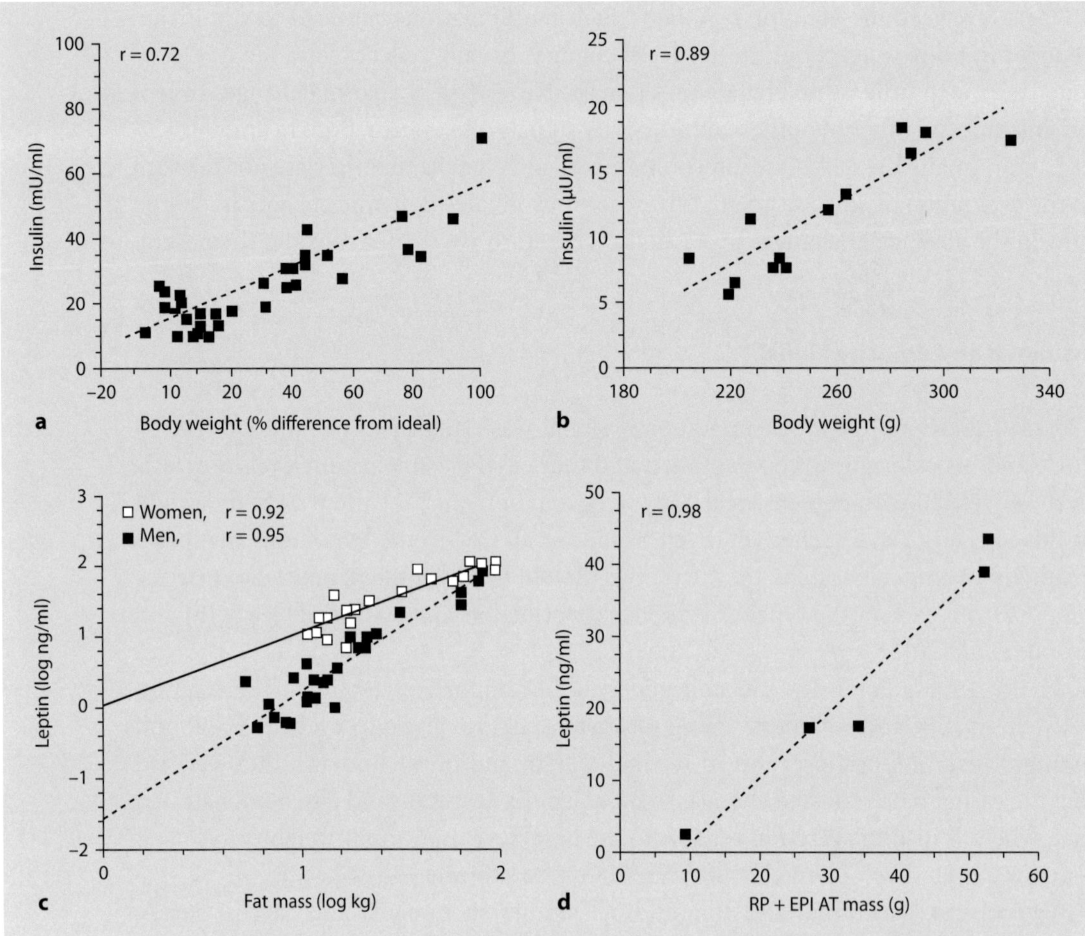

Fig. 2. Relationships between basal plasma levels of insulin and leptin and adiposity. **a** Basal insulin levels versus ideal body weight in humans. Modified with permission from Bagdade et al. [31]. **b** Basal insulin levels versus body weight in rats that were fed ad lib, intragastrically overfed to 125% normal weight, or restrictively fed to 80% normal weight. Modified with permission from Bernstein et al. [32]. **c** Basal leptin levels versus fat mass in lean and obese men and women. Modified with permission from Rosenbaum et al. [46]. **d** Basal leptin levels in rats versus the sum of the weights of the retroperitonal (RP) and epididymal (EPI) fat pads (AT). Each point plotted is the mean value of one of five groups of rats: from left to right, 'pre-obese' rats i.e., rats fed a low-fat diet for 16 weeks; rats fed a high-fat diet for 16 weeks and then restrictively fed; rats fed a high-fat diet for 16 weeks, then restrictively fed, then allowed ad lib access to low-fat diet; and rats a fed high-fat diet for 16 weeks. Data used with permission from MacLean et al. [36].

perturbations. Basal plasma insulin, however, returned to or below control levels well before AT mass was corrected (fig. 3), indicating that insulin did not encode AT mass accurately through most of the period of compensation and could not have driven the regulatory responses correcting AT mass. Similar data have been reported [38, 39].

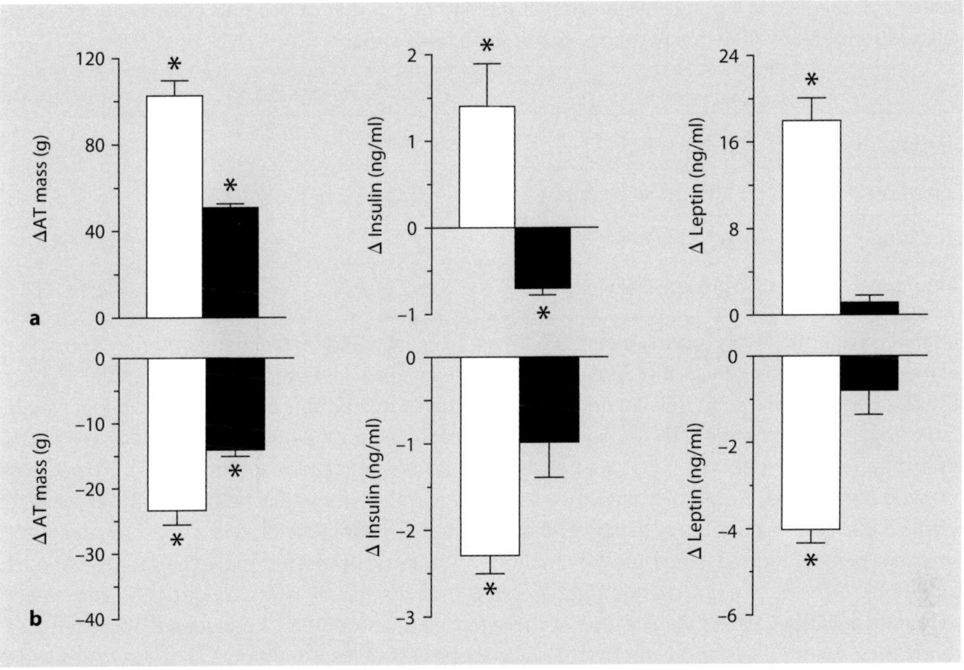

Fig. 3. Relationships between adipose tissue (AT) mass and basal plasma levels of insulin and leptin in rats during dynamic regulation of AT mass following (**a**) forced weight gain produced by intragastric overfeeding or (**b**) forced weight loss produced by restrictive feeding. Data are differences from age-matched ad lib-fed control rats, sampled on the last day of weight manipulation (open bars) or after 1 (**b**) or 2 (**a**) weeks of compensation (black bars). Note scale changes in the different panels. AT mass was estimated using computed tomography. Blood samples were taken via chronic jugular vein cannulas late in the diurnal phase after 3 h food deprivation. Note (1) that overfed rats' basal plasma insulin and leptin were no longer higher than control rats' after 2 wk of compensation despite that overfed rats still had 51 g excess AT, and (2) that underfed rats' basal plasma insulin and leptin were no longer different from control rats' after 1 week of compensation despite that underfed rats still had a 14 g deficit in AT. From Gloy et al. [unpubl. data]. *Significant difference between control and test rats, $p < 0.001$.

Leptin as an Adiposity Signal

Leptin has been considered an adiposity signal since its discovery as a secretory product of the adipocyte [40]. Although leptin secretion is affected by eating, fasting, the metabolic state of the adipocyte and a variety of other factors [41–43], basal leptin secretion is closely associated with AT mass [44–46]. Interestingly, for any level of AT, basal leptin levels are higher in women than men and higher in pre- than postmenopausal women [46, 47] (fig. 2c).

Basal leptin levels have also been associated with AT mass in normal-weight and obese humans before and after weight gain or loss by overfeeding or dieting [44, 46, 48–51]. These studies indicate that weight gains or losses of up to ~20% increase or

Table 1. Basal plasma leptin levels (ng/ml) at the end of a period of dynamic body-weight change versus during steady-state maintenance of the same body weight

		Dynamic weight change	Steady state
Condition 1	at initial weight	48.1±7.9*	82.7±6.4
Condition 2	at 90% initial weight	23.2±3.8*	45.4±9.2
Condition 3	at 80% initial weight	17.7±4.4*	29.6±6.4
Condition 4	at 110% initial weight	82.9±11.4	81.1±10.7

Basal plasma leptin levels (ng/ml; mean ± SEM) are from lean and obese men and pre-menopausal women who entered a clinical metabolism unit at their maximal lifetime body weight (initial weight). After >2 weeks stabilization on a liquid formula diet given in amounts adjusted to maintain weight (<10 g/day weight change over 14 days), some subjects received 5,000–8,000 kcal/day solid food until they reached 110% of initial weight and then were re-stabilized (Condition 4). They then received 800 kcal/day liquid diet until they returned to initial weight (Condition 1). Other subjects received 800 kcal/day until they reached 90 or 80% initial weight (Conditions 2 and 3, respectively; note that percent fat mass, measured by hydrodensitometry, decreased more than weight in each condition). 'Dynamic weight change' data are on the day of reaching the weight goal, and 'steady state' data are after 2 weeks stabilization at the same weight. Condition 1 included 8 obese females; condition 2, 1 non-obese and 2 obese males and 7 obese females; condition 3, 2 obese males and 9 obese females; condition 4, 2 nonobese and 2 obese males and 1 nonobese and 8 obese females. Data used with permission from Rosenbaum et al. [46].
*Significant difference, $p < 0.005$.

decrease, respectively, basal plasma leptin levels in both lean and obese subjects [44, 46, 48–51]. Korner et al. [52] recently extended this to even larger weight losses, up to 40%, in bariatric surgery patients. All these data are consistent with the hypothesis that leptin is an adiposity signal. In some situations, however, the relationship of basal leptin to AT mass is altered, indicating that basal leptin does not simply encode AT mass. For example, basal leptin levels in Roux-en-Y gastric bypass patients at ≥1 year follow-up were only ~40% of those in BMI-matched control women [37]. Similarly, Rosenbaum et al. [46] reported that the ratio of log plasma leptin to log fat mass (FM; measured by hydrodensitometry) was decreased in women, but not men, who were maintained at a stable lower body weight, whereas log leptin:log FM ratio was increased in men, but not women, who were maintained at a stable higher weight. In addition, in the same study, leptin levels at a stable level of weight were markedly lower if the subjects had been in chronic negative energy balance than after 2 weeks of stable maintenance at the same weight (table 1, conditions 1–3). This indicates that the predominate effect of negative energy balance on basal plasma leptin levels is maintained for weeks of dieting and, therefore, that plasma leptin levels can not encode AT mass under these conditions because it can not distinguish a dynamic state of negative energy balance from adiposity. Interestingly, leptin levels during positive energy balance and after 2 weeks in energy balance at the same weight were not different (table 1, condition 4); thus, leptin could encode AT mass accurately during weight gain.

Basal leptin levels have also been associated with AT mass in rodents [36, 53, 54]. In the MacLean et al. [36] study described above, there was also a positive correlation between basal plasma leptin levels and AT mass in rats whose body weights had been increased or decreased (fig. 2d). In many rodent studies, however, body weight rather than AT mass is measured. This is an important source of error because the relationship between body weight and AT mass is highly variable in rodents, especially after weight has been manipulated. Thus, the relations between leptin levels and body weight may appear quite different than the relations between leptin levels and AT mass [36, 54]. For example, in sub-strains of rats that are more (diet-induced obese; DIO) or less (diet-resistant; DR) susceptible to diet-induced obesity, equal degrees of weight loss produced markedly larger falls in basal plasma leptin and insulin in DIO rats than DR rats [54]. Loss of AT mass also differed between the strains, however, so that the leptin:AT and insulin:AT ratios remained relatively constant. Our study of dynamic regulation of AT mass also indicates that leptin is insufficient to signal adiposity (fig. 3). Rather, basal plasma leptin was not significantly different from control levels 1–2 weeks after allowing rats to correct AT perturbations despite that AT mass had only begun to normalize (fig. 3). These data indicate that basal plasma leptin does not encode information about AT mass that drives regulatory responses in either underweight or overweight rats.

Conclusions

A substantial database clearly indicates that AT mass is highly positively correlated with basal plasma insulin and leptin levels in both animal and human subjects at steady states of AT mass, as shown in figure 1b. This supports the endocrine coding criterion for the hypothesized roles of insulin and leptin as adiposity signals. We emphasize, however, that this conclusion pertains only when AT mass is constant. This qualification is rarely mentioned in the literature, which gives the impression that there is more support for the endocrine coding criterion than actually exists. Indeed, the few extant data related to dynamic states of AT mass provide little support for the hypotheses. Therefore, our conclusions, based on the premise that evidence of appropriate changes in hypothesized adiposity signals in situations in which the organism is dynamically regulating AT mass is required to establish the endocrine coding criterion and the fact that such data are scarce and inconsistent, are: (1) the cases for insulin and leptin as adiposity signals are overstated, (2) how adiposity is signaled to the brain remains an open question in the physiology of energy homeostasis, and (3) further tests of the hypotheses that insulin and leptin, as well as amylin, ghrelin and PYY, are adiposity signals are urgently needed.

An important but unresolved issue facing future work in this area is that we do not know what parameters of hormone levels encode adiposity. Basal levels and integrated daily averages have face validity as tonic signals, but it is entirely possible that the central nervous system reacts to other parameters, such as levels at certain times of day or changes after meals. Guo et al.'s [55] study of PYY levels in lean and obese humans is illustrative. BMI was significantly negatively correlated with basal PYY level, but not with either the area under the curve (AUC) of PYY levels during 3 h after a test meal or the peak postprandial PYY level. When weight change was measured one year later, however, peak postprandial PYY level, but neither basal nor postprandial AUC PYY level, correlated significantly with weight change, suggesting that this parameter, not basal PYY level, was an effective tonic signal.

Our review, like most of the literature, has focused on the idea that there is a single adiposity signal. Instead, perhaps the signal is compound. For example, if the brain monitors both basal insulin and leptin levels and whether energy balance is positive or negative, it could resolve the apparently incorrect encoding of AT mass by insulin and leptin in Roux-en-Y patients described above [52]. A related possibility is that adiposity signals are situationally specific, i.e. different adiposity signals are prepotent in different physiological situations. Because both insulin and leptin are pleiotropic hormones, it is to be expected that numerous other factors can influence their basal levels. For example, perhaps basal insulin is disproportionately decreased in Roux-en-Y patients [52] because of their exaggerated incretin secretion, which leads to more rapid clearance of prandial increases in blood glucose and overall greater insulin sensitivity. In view of the many influences on basal insulin and leptin levels, it seems unreasonably parsimonious to expect that either could consistently encode AT mass accurately. Finally, that neurons in the hypothalamic arcuate nucleus have receptors for leptin, insulin and ghrelin and also are sensitive to several metabolites [5, 56] is also consistent with the idea that AT mass is encoded by a compound signal.

References

1 Friedman JM: Obesity: causes and control of excess body fat. Nature 2009;459:340–342.

2 Keesey RE, Powley TL: The regulation of body weight. Annu Rev Psychol 1986;37:109–133.

3 Keesey RE, Powley TL: Body energy homeostasis. Appetite 2008;51:442–445.

4 Kennedy GC: The role of depot fat in the hypothalamic control of food intake in the rat. Proc R Soc Lond B Biol Sci 1953;140:578–596.

5 Langhans W, Geary N: Overview of the physiological control of eating; in Langhans W, Geary N (eds): Frontiers in Eating and Weight Regulation. Forum Nutr. Basel, Karger, 2010, vol 63, pp 9–53.

6 Leibel RL: Molecular physiology of weight regulation in mice and humans. Int J Obes (Lond) 2008; 32(suppl 7):S98–S108.

7 Morton GJ, Cummings DE, Baskin DG, Barsh GS, Schwartz MW: Central nervous system control of food intake and body weight. Nature 2006;443:289–295.

8 Schwartz MW, Woods SC, Porte Jr D, Seeley RJ, Baskin DG: Central nervous system control of food intake. Nature 2000;404:661–671.

9 Schwartz MW, Woods SC, Seeley RJ, Barsh GS, Baskin DG, Leibel RL: Is the energy homeostasis system inherently biased toward weight gain? Diabetes 2003;52:232–238.

10 Gallagher D, Visser M, Sepulveda D, Pierson RN, Harris T, Heymsfield SB: How useful is body mass index for comparison of body fatness across age, sex, and ethnic groups? Am J Epidemiol 1996;143: 228–239.

11 Hillebrand JJ, Langhans W, Geary N: Validation of computed tomographic estimates of intra-abdominal and subcutaneous adipose tissue in rats and mice. Obesity 2009;in press.
12 Levin BE, Dunn-Meynell AA: Defense of body weight against chronic caloric restriction in obesity-prone and -resistant rats. Am J Physiol Regul Integr Comp Physiol 2000;278:R231–R237.
13 Wirtshafter D, Davis JD: Set points, settling points, and the control of body weight. Physiol Behav 1977;19:75–78.
14 Keesey RE, Hirvonen MD: Body weight set-points: determination and adjustment. J Nutr 1997;127: 1875S–1883S.
15 Jobst EE, Enriori PJ, Cowley MA: The electrophysiology of feeding circuits. Trends Endocrinol Metab 2004;15:488–499.
16 Blevins JE, Baskin DG: Hypothalamic-Brainstem Circuits Controlling Eating; in Langhans W, Geary N (eds): Frontiers in Eating and Weight Regulation. Forum Nutr. Basel, Karger, 2010, vol 63, pp 133–140.
17 Coleman DL: Obese and diabetes: two mutant genes causing diabetes-obesity syndromes in mice. Diabetologia 1978;14:141–148.
18 Harris RB, Martin RJ: Specific depletion of body fat in parabiotic partners of tube-fed obese rats. Am J Physiol 1984;247:R380–R386.
19 Davis JD, Wirtshafter D, Asin KE, Brief D: Sustained intracerebroventricular infusion of brain fuels reduces body weight and food intake in rats. Science 1981;212:81–83.
20 Woods SC, Lutz TA, Geary N, Langhans W: Pancreatic signals controlling food intake; insulin, glucagon and amylin. Philos Trans R Soc Lond B Biol Sci 2006;361:1219–1235.
21 Lutz TA: Roles of amylin in satiation, adiposity and brain development; in Langhans W, Geary N (eds): Frontiers in Eating and Weight Regulation. Forum Nutr. Basel, Karger, 2010, vol 63, pp 64–74.
22 Cummings DE, Overduin J: Gastrointestinal regulation of food intake. J Clin Invest 2007;117:13–23.
23 Geary N: Endocrine controls of eating: CCK, leptin, and ghrelin. Physiol Behav 2004;81:719–733.
24 Neary MT, Batterham RL: Peptide YY: food for thought. Physiol Behav 2009;97:616–619.
25 Cone RD, Low MJ, Elmquist JK, Cameron JL: Neuroendocrinology; in Larsen PR, Kronenberg HM, Melmed S, Polonsky KS (eds): Williams Textbook of Endocrinology. Philadelphia, Saunders, 2009, vol 10, pp 81–176.
26 Banks WA: Blood-brain barrier as a regulatory interface; in Langhans W, Greary N (eds): Frontiers in Eating and Weight Regulation. Forum Nutr. Basel, Karger, 2010, vol 63, pp 102–110.
27 Levin BE, Dunn-Meynell AA, Ricci MR, Cummings DE: Abnormalities of leptin and ghrelin regulation in obesity-prone juvenile rats. Am J Physiol Endocrinol Metab 2003;285:E949–E957.
28 Ahima RS, Prabakaran D, Mantzoros C, et al: Role of leptin in the neuroendocrine response to fasting. Nature 1996;382:250–252.
29 Cooper JA, Polonsky KS, Schoeller DA: Serum leptin levels in obese males during over- and underfeeding. Obesity (Silver Spring) 2009;in press.
30 Challis BG, Coll AP, Yeo GS, et al: Mice lacking pro-opiomelanocortin are sensitive to high-fat feeding but respond normally to the acute anorectic effects of peptide-YY(3–36). Proc Natl Acad Sci USA 2004;101:4695–4700.
31 Bagdade JD, Bierman EL, Porte Jr D: The significance of basal insulin levels in the evaluation of the insulin response to glucose in diabetic and nondiabetic subjects. J Clin Invest 1967;46:1549–1557.
32 Bernstein IL, Lotter EC, Kulkosky PJ, Porte Jr D, Woods SC: Effect of force-feeding upon basal insulin levels of rats. Proc Soc Exp Biol Med 1975; 150:546–548.
33 Woods SC, Decke E, Vasselli JR: Metabolic hormones and regulation of body weight. Psychol Rev 1974;81:26–43.
34 Polonsky KS, Given BD, Van Cauter E: Twenty-four-hour profiles and pulsatile patterns of insulin secretion in normal and obese subjects. J Clin Invest 1988;81:442–448.
35 Polonsky KS, Given BD, Hirsch L, et al: Quantitative study of insulin secretion and clearance in normal and obese subjects. J Clin Invest 1988;81:435–441.
36 MacLean PS, Higgins JA, Jackman MR, et al: Peripheral metabolic responses to prolonged weight reduction that promote rapid, efficient regain in obesity-prone rats. Am J Physiol Regul Integr Comp Physiol 2006;290:R1577–R1588.
37 Korner J, Bessler M, Cirilo LJ, et al: Effects of Roux-en-Y gastric bypass surgery on fasting and postprandial concentrations of plasma ghrelin, peptide YY, and insulin. J Clin Endocrinol Metab 2005;90: 359–365.
38 Almeida NG, Levitsky DA, Strupp B: Enhanced thermogenesis during recovery from diet-induced weight gain in the rat. Am J Physiol 1996;271: R1380–R1387.
39 Harris RB, Kasser TR, Martin RJ: Dynamics of recovery of body composition after overfeeding, food restriction or starvation of mature female rats. J Nutr 1986;116:2536–2546.
40 Zhang Y, Proenca R, Maffei M, Barone M, Leopold L, Friedman JM: Positional cloning of the mouse obese gene and its human homologue. Nature 1994; 372:425–432.

41 Cammisotto PG, Gelinas Y, Deshaies Y, Bukowiecki LJ: Regulation of leptin secretion from white adipocytes by insulin, glycolytic substrates, and amino acids. Am J Physiol Endocrinol Metab 2005;289: E166–E171.

42 Lee MJ, Fried SK: Integration of hormonal and nutrient signals that regulate leptin synthesis and secretion. Am J Physiol Endocrinol Metab 2009;296: E1230–E1238.

43 Weigle DS, Duell PB, Connor WE, Steiner RA, Soules MR, Kuijper JL: Effect of fasting, refeeding, and dietary fat restriction on plasma leptin levels. J Clin Endocrinol Metab 1997;82:561–565.

44 Considine RV, Sinha MK, Heiman ML, et al: Serum immunoreactive-leptin concentrations in normal-weight and obese humans. N Engl J Med 1996;334: 292–295.

45 Maffei M, Halaas J, Ravussin E, et al: Leptin levels in human and rodent: measurement of plasma leptin and ob RNA in obese and weight-reduced subjects. Nat Med 1995;1:1155–1161.

46 Rosenbaum M, Nicolson M, Hirsch J, Murphy E, Chu F, Leibel RL: Effects of weight change on plasma leptin concentrations and energy expenditure. J Clin Endocrinol Metab 1997;82:3647–3654.

47 Saad MF, Damani S, Gingerich RL, et al: Sexual dimorphism in plasma leptin concentration. J Clin Endocrinol Metab 1997;82:579–584.

48 Wing RR, Sinha MK, Considine RV, Lang W, Caro JF: Relationship between weight loss maintenance and changes in serum leptin levels. Horm Metab Res 1996;28:698–703.

49 Rosenbaum M, Goldsmith R, Bloomfield D, et al: Low-dose leptin reverses skeletal muscle, autonomic, and neuroendocrine adaptations to maintenance of reduced weight. J Clin Invest 2005;115: 3579–3586.

50 Levine JA, Eberhardt NL, Jensen MD: Leptin responses to overfeeding: relationship with body fat and nonexercise activity thermogenesis. J Clin Endocrinol Metab 1999;84:2751–2754.

51 Kolaczynski JW, Ohannesian JP, Considine RV, Marco CC, Caro JF: Response of leptin to short-term and prolonged overfeeding in humans. J Clin Endocrinol Metab 1996;81:4162–4165.

52 Korner J, Inabnet W, Febres G, et al: Prospective study of gut hormone and metabolic changes after adjustable gastric banding and Roux-en-Y gastric bypass. Int J Obes (Lond) 2009;33:786–795.

53 Frederich RC, Hamann A, Anderson S, Lollmann B, Lowell BB, Flier JS: Leptin levels reflect body lipid content in mice: evidence for diet-induced resistance to leptin action. Nat Med 1995;1:1311–1314.

54 Levin BE, Keesey RE: Defense of differing body weight set points in diet-induced obese and resistant rats. Am J Physiol 1998;274:R412–R419.

55 Guo Y, Ma L, Enriori PJ, et al: Physiological evidence for the involvement of peptide YY in the regulation of energy homeostasis in humans. Obesity (Silver Spring) 2006;14:1562–1570.

56 Moran TH: Hypothalamic nutrient sensing and energy balance; in Langhans W, Geary N (eds): Frontiers in Eating and Weight Regulation. Forum Nutr. Basel, Karger, 2010, vol 63, pp 94–101.

Jacquelien J.G. Hillebrand
ETH Zürich
Schorenstrasse 16
CH–8603 Schwerzenbach (Switzerland)
Tel. +41 44 655 7390, Fax +41 44 655 7206, E-Mail jacquelien-hillebrand@ethz.ch

Langhans W, Geary N (eds): Frontiers in Eating and Weight Regulation.
Forum Nutr. Basel, Karger, 2010, vol 63, pp 123–132

Leptin-Signaling Pathways and Leptin Resistance

Heike Münzberg

Pennington Biomedical Research Center, LSU System, Baton Rouge, La., USA

Abstract

Leptin acts as an anorexigenic hormone in the brain, where the long form of the leptin receptor (LRb) is widely expressed in hypothalamic and extra-hypothalamic sites that are known to participate in diverse feeding circuits. The important role of leptin in energy homeostasis is demonstrated by the profound hyperphagia and morbid obesity in humans and rodents null for leptin or LRb. However, common forms of obesity are associated with high leptin levels and a failure to respond effectively to exogenous leptin; indicating a state of leptin resistance. Leptin resistance is thought to be an important component in the development of obesity. Several defects may contribute to the leptin resistant state, including a defective leptin transport across the blood-brain barrier, which reduces the availability of leptin at its receptor. Furthermore, defects in LRb signal transduction involving reduced LRb expression or the induction of feedback inhibitors have been found in leptin resistance; these defects are commonly termed cellular leptin resistance,. Finally, reduced leptin action can result in the disruption of proper neuronal interactions, by altering neuronal wiring. Interestingly, some leptin functions remain intact in the leptin-resistant state, such as cardiovascular leptin effects. The appearance of selective leptin resistance is mirrored by the observation that cellular leptin resistance has been found only in some subpopulations of hypothalamic LRb neurons. Current efforts to dissect leptin function in specific populations of LRb neurons will increase our understanding of these complexities of leptin physiology.

Leptin Action in the Central Nervous System

Leptin Receptor

The adipocyte-derived hormone leptin executes its anorexigenic actions via leptin receptors (LR), which belong to the category of class I cytokine receptors. Six LR variants (LRa-f in the mouse) exist, all derived from alternative splicing of the *lepr* gene. LR variants are subcategorized as short form (LRa, c, d, f, g), soluble form (LRe) and long form (LRb). LRa and LRb consist of identical extracellular domains,

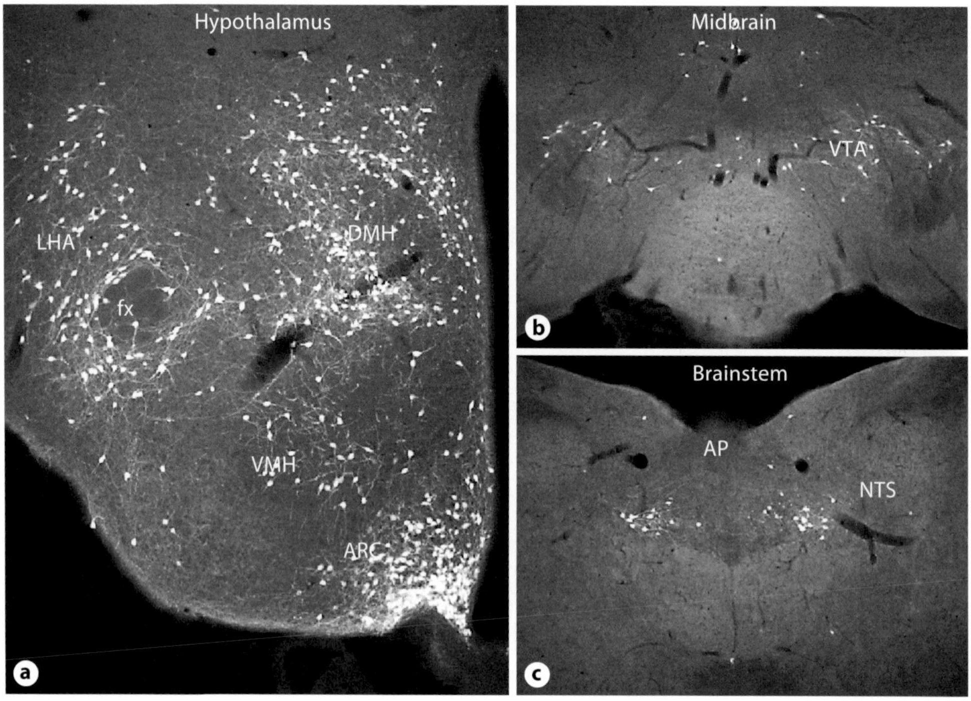

Fig. 1. Immunohistochemical visualization of green fluorescent protein expressing LRb neurons in the hypothalamus (**a**), midbrain (**b**) and brainstem (**c**). LRb neurons are found in central sites that contribute importantly to the control of eating. LHA = Lateral hypothalamic area; fx = fornix; DMH = dorsomedial hypothalamus; VMH = ventromedial hypothalamus; Arc = arcuate nucleus; VTA = ventral tegmental area; AP = area postrema; NTS = nucleus of the solitary tract.

transmembrane domains and the initial 29 amino acids of the intracellular domain, but only LRb possesses a long intracellular signaling domain [1, 2]. Even though the function of short and soluble LR is still unclear, LRb appears to be the critical receptor for leptin action, as the dramatic obese and hyperphagic phenotype of *lepr^{db}* mice is due to a mutation that prevents the protein expression of LRb, whereas all other LR variants are still functional [2]. Also, *lepr^{db}* mice are phenotypically identical to leptin-deficient *lep^{ob}* mice as well as to mice with deletion of all LR isoforms [3].

LRb is predominantly expressed in the CNS, and central LRb expression has been found sufficient to normalize the phenotype of *lepr^{db}* mice [3]. LRb expression is particularly high in the hypothalamus, where diverse sensory information is integrated and processed to adjust homeostatic function, as discussed in several chapter in this volume[4, 5]. Apart from its obvious effects on body weight and food intake, leptin has also been found to control thyroid axis, reproductive axis, glucose homeostasis, immune function, growth and autonomic nervous system and other physiological functions [6].

Leptin Action in AgRP and POMC Neurons in the Arcuate Nucleus

The hypothalamic arcuate nucleus (Arc), which stretches along the base of the hypothalamus, is so far the best-characterized site of leptin action. At least two distinct neuronal populations in the Arc express LRb: the orexigenic agouti-related peptide (AgRP) neurons and the anorexigenic pro-opiomelanocortin (POMC)-expressing neurons [7]. Leptin stimulates the expression and release of the POMC-derived peptide α-melanocyte-stimulating hormone (α-MSH), which in turn activate CNS melanocortin receptors (MC3R and MC4R) on target neurons. In contrast, leptin reduces AgRP expression, which inhibits of MC3R and MC4R signaling [8, 9]. POMC- and AgRP- expressing neurons densely innervate the paraventricular hypothalamic nucleus (PVN), the dorsomedial hypothalamus, and the lateral hypothalamic area and other sites [10, 11] ultimately leading to neuroendocrine, autonomic and behavioral outputs [12]. Deletion of MC4R leads to morbid obesity [8]. Thus, it has been hypothesized that leptin action on Arc melanocortin neurons is the key mechanism for leptin-regulated energy homeostasis.

Leptin Action Outside of the Arc

Several sites in the CNS express LRb, e.g. the hypothalamus, midbrain and brainstem (fig. 1) [13]. Most of these LRb-expressing sites are known to regulate energy homeostasis, suggesting that anorexigenic leptin actions involve several components. Nevertheless, the functions of LRb neurons outside the Arc are not yet fully understood. While LRb-expressing Arc neurons are important for the response to leptin, it has become clear from recent literature that they may not account for all or even the majority of anorexigenic leptin action in vivo. For example, LRb neurons in the ventromedial hypothalamus (VMH) [14], and the ventral tegmental area (VTA) [15, 16] have been shown to participate in the regulation of eating and energy balance, whereas LRb neurons in the premammillary nucleus do not participate in eating-related circuits, but are involved in reproductive function [17]. In contrast, reactivation of LRb specifically in the Arc of LRb-deficient mice surprisingly reduced body weight only modestly, but normalized glucose homeostasis and locomotor activity [18]. Furthermore, deletion of the leptin receptor gene *(Lepr)* from AgRP and/or POMC neurons results in modest obesity compared to that of $lepr^{db}$ animals [19, 20]. Hence, while deletion of LRb from specific neuronal populations in the ARC, VMH or the VTA clearly demonstrate that leptin has important actions in these sites, the dramatic obesity and hyperphagia found in leptin-signaling deficient lep^{ob} or $lepr^{db}$ mice could not be recapitulated. Therefore, despite our increased understanding of central leptin action, many details still await discovery and further efforts are on the way to define specific subpopulation of LRb neurons that contribute to anorexigenic leptin actions.

Leptin Signaling Pathways

LRb is a typical class I cytokine receptor without intrinsic kinase activity to initiate the phosphorylation of tyrosin residues (which are common stimulatory events to induce signaling pathways). Therefore, LRb binds janus kinase 2 (JAK2) for the induction of signaling pathways. Leptin binding to LRb results in conformational changes of LRb that enables binding and autophosphorylation of JAK2. JAK2 promotes the phosphorylation of three conserved tyrosine residues (Y) in the LRb signaling domain (Y985, Y1077, Y1138). Phosphorylated tyrosine residues (PY) are common docking stations for adaptor molecules that interact with PY via so called SH-2 domains. In cell culture experiments these three tyrosine residues on LRb have been found to interact with distinct adaptor proteins from discrete signaling pathways [21]. Thus, together with tyrosine phosphorylated JAK2, there are 4 PY sites that induce specific downstream LRb signaling events (fig. 2).

PY985 interacts with tyrosine phosphatase SHP-2, which leads to the activation of the ERK signaling pathway. PY1077 binds the nuclear transcription factor signal-transducer-and-activator-5 (STAT5), which is then phosphorylated by JAK2, dimerized, and translocated into the nucleus, where STAT5 acts as a transcription factor. Similarly, PY1138 binds specifically to STAT3, which is also phosphorylated, dimerized, and transported into the nucleus for transcriptional control [21]. The majority of LRb/STAT induced transcripts are unknown, although leptin-induced transcription of the neuropeptide POMC as well as the suppressor-of-cytokine-signaling-3 (SOCS3) has been demonstrated to require LRb/STAT3 signaling [22, 23]. SOCS-3 acts as a negative feedback signal on LRb and has been shown to inhibit JAK2 phosphorylation by interacting with PY985 as well as directly via JAK2 [24].

Most of these interactions of LRb with specific signaling pathways are based on cell-culture models, but some physiological roles for discrete LRb signaling pathways have been dissected. For example, the majority of leptin's anorexigenic action can be attributed to PY1138, because mice with targeted mutation of Y1138, which leads to a loss of LRb/STAT3 binding without affecting other LRb signaling pathways, are severely obese and hyperphagic, comparable to *lepr^db^* mice [25]. To the contrary, targeted mutation of Y938, which leads to a selective loss of LRb/SHP-2 binding, results in lean mice with increased leptin sensitivity, suggesting that an inhibitory component of the LRb signal has been lost in these mice [26]. Therefore, the results from Y938-mutated mice support an important role of this phosphorylation site for the binding and function of the negative-feedback mediator SOCS-3, as suggested by earlier cell-culture experiments [24].

Leptin also has been found to regulate several other pathways involved in the regulation of energy homeostasis, including the mammalian target of rapamycin (mTOR), AMP-activated protein kinase (AMPK), and the IRS/PI3K pathway [27–29]. However, the exact molecular mechanisms resulting in their interaction with the leptin signaling pathway, and whether leptin indeed directly regulates these pathways remains

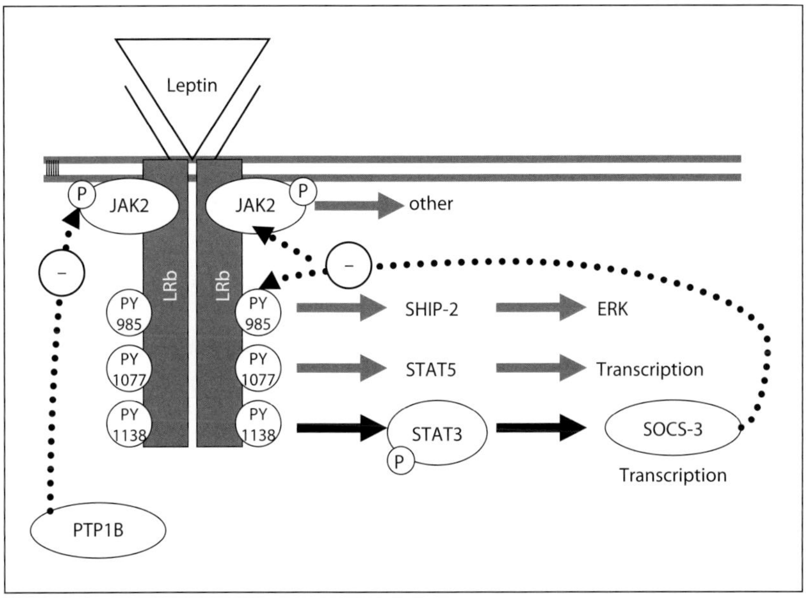

Fig. 2. The intracellular LRb signaling cascade involves the binding, phosphorylation and activation of JAK2 (janus kinase 2) as well as phosphorylation of conserved tyrosine residues on LRb. These initiate distinct signaling pathways to mediate the diverse functions of leptin, including the transcriptional regulation of SOCS-3 (suppressor-of-cytokine-signaling-3) and PTP1B (phosphotyrosine phosphatase 1B) which are negative regulator or LRb signaling and are involved in the development of leptin resistance in the obese state.

unknown. Furthermore, leptin regulates cellular function on multiple levels involving – aside from signaling pathways and transcriptional events – the stimulation or inhibition of neuronal activity (leading to neuropeptide and transmitter release), neuronal plasticity via axonal outgrowth, and reorganization as well as neurogenesis [9, 10, 30–32].

Leptin Resistance

Despite the dramatic effects of leptin deficiency on body weight and food intake in leptin-deficient mice, the use of leptin as an anorexigenic drug to treat human obesity was disappointing [33]. Because obesity is generally associated with elevated leptin levels [34] it was suggested that obesity causes a decrease in leptin efficiency. The reduced potency of leptin to inhibit eating has been termed leptin resistance.

Leptin resistance can be dramatically demonstrated in rodents in which obesity is induced by feeding a palatable, high-fat diet (this condition has been termed diet-induced obesity, DIO). Such mice become progressively less sensitive to leptin administration [35]. Leptin resistance has been also described in aged individuals [36], seasonal obese rodents [37] and DIO prone (compared to DIO resistant) rodent strains [38], which all share a state of increased adiposity and elevated leptin levels. Whereas the concept of leptin resistance is well accepted, its mechanisms remain incompletely understood. Leptin transport defects [39], defects within the LRb signaling cascade (cellular leptin resistance) [35, 40], and defective neuronal wiring and function of eating circuits [41, 42] all may contribute.

Defective Leptin Transport across the Blood-Brain Barrier

Leptin is transported across the blood-brain barrier into the brain by a saturable, regulated transporter [43, 44]. In DIO rats the ratio of cerebrospinal fluid (CSF) leptin versus serum leptin is decreased. Thus, even though CSF leptin levels in DIO rats is still elevated compared to lean control rats, the central leptin availability at LRb might not be sufficient to translate into appropriate LRb signaling action [39]. Consistent with this model is the fact that DIO-induced leptin resistance can be improved (but not completely recovered) by central leptin injections, which circumvent any defective transport mechanisms [35]. Thus, defective leptin transport represents one component in the leptin resistant state.

Cellular Leptin Resistance and Negative Feedback Signals

Cellular leptin resistance refers to any decrease in intracellular LRb signaling events. Indeed, DIO can be also characterized by decreased leptin-induced STAT3 activation [35]. Furthermore, negative-feedback mechanisms have been described for LRb in vivo via SOCS-3 and protein tyrosine phosphatase PTP1B.

SOCS-3 is upregulated in response to leptin within hypothalamic areas known to express LRb [22]. Thus, increased circulating leptin, as found in most obese individuals, may chronically induce SOCS-3 levels and therefore attenuate leptin signaling via LRb. Recent data demonstrated that targeted deletion of the SOCS-3 gene as well as conditional, neuron-specific SOCS-3 deletion resulted in reduced weight gain on a high fat diet and increased leptin sensitivity [45], thus supporting that SOCS-3 indeed interacts with LRb signaling in vivo and importantly mediates leptin resistance.

PTB1B is a cytoplasmic protein that, like SOCS-3, has been found to decrease JAK2 phosphorylation and also decreases LRb signaling [46]. Similar to SOCS-3 the complete deletion of PTP1B or conditional PTB1B deletion in neurons resulted in decreased body weight gain on a high-fat diet and a lean phenotype [47]. Leptin seems to increase PTB1B via unknown mechanisms [47, 48]. Thus, an important open question is whether PTB1B and SOCS-3 are independently regulated by different mechanisms.

It is also possible that other leptin-independent factors may be involved in the regulation of PTB1B or SOCS-3. For example, both have been shown to be upregulated by high-fat diet feeding [40, 48] and SOCS-3 is also regulated by several cytokines that are known to be induced in obese subjects [49]. Thus, SOCS-3 and PTB1B seem to have important roles in the development of leptin resistance and, therefore, are potential targets for the development of anti-obesity drugs.

Leptin-Induced Leptin Resistance

Leptin resistance is commonly associated with hyperleptinemia, and leptin can induce SOCS-3 as well as PTB1B, suggesting that leptin causes its own resistance. This concept was further supported in rats receiving prolonged central leptin infusion: Even though these animals initially lose weight, their body weights normalize over time despite ongoing leptin treatment. In this state even additional acute leptin injections failed to induce any anorexigenic effect [50–52]. Indeed, when put on a high-fat diet, animals with acquired leptin resistance gained more weight than control animals, further indicating that leptin resistance contributes significantly to DIO.

In contrast, low leptin levels (e.g. in mice after prolonged fasting) as well as lack of leptin (lep^{ob} mice) increases leptin sensitivity, consistent with the idea that low leptin levels result in decreased SOCS-3 levels. Therefore, leptin induced STAT3 activation (stimulating SOCS-3 expression) and SOCS-3 induction (inhibiting leptin induced STAT3 activation) might normally be in a steady state for any given level of circulating leptin [6]. The evolutionary adaptiveness of a leptin-driven feedback mechanism in which low body weight (low leptin levels) produces the most potent anorexigenic effect seems questionable. A potential explanation for this is that leptin action may be less important in states of plenty, but be critical to coordinate reproductive behavior with information about body fat stores in states of scarcity.

Site-Specific Leptin Resistance

Although leptin's anorexigenic action is reduced in leptin resistant states, other leptin-regulated processes have been shown to be preserved. For example, leptin's effect to increase sympathetic tone in the kidney, resulting in increased blood pressure, is preserved in leptin-resistant mice [53]. This selective leptin resistance cannot be easily explained by a defect in leptin transport or general cellular signaling mechanisms. Rather, this suggests that only subpopulations of LRb neurons are involved in either central eating-control circuits or cardiovascular circuits and that these LRb circuits are selectively affected during leptin resistance.

Indeed, in DIO mice, cellular leptin resistance with reduced leptin-induced STAT3 phophorylation and increased SOCS-3 was found to be restricted to the Arc, with no such resistance evident in other hypothalamic and extrahypothalamic sites [40]. Similarly, SOCS-3 levels were selectively elevated in the Arc in a seasonal mammal during the period of increased adiposity and hyperphagia [37], further supporting the existence of site-specific leptin resistance in the brain. Many LRb neurons in the Arc are neuroendocrine cells, with processes exposed to the portal vascular system. These cells have been shown to respond more sensitively and faster to changes in circulating leptin levels, suggesting that leptin could directly stimulate these neurons in addition to a leptin transport across the blood-brain barrier [47]. Therefore, increased leptin access to arcuate LRb neurons would make them more prone to the development of increased SOCS-3 levels and cellular leptin resistance [54].

Site-specific cellular leptin resistance has been also demonstrated in pregnant, hyperphagic rats, although in this model leptin resistance was confined to the VMH [55]. Many VMH neurons co-express LRb and estrogen receptors [56], so that the substantial changes in levels of estrogens during the ovarian cycle and pregnancy might interact with LRb signaling in the VMH. Thus, the leptin resistant state might be a highly regulated response to diverse metabolic situations, suggesting that we are only beginning to understand the many facets that contribute to the development of leptin resistance and obesity.

Conclusion

The research initiated by the discovery of leptin and its receptor, together with advances in novel genetic and histological methodologies, has tremendously increased our knowledge about central circuits that control energy homeostasis. Future research aimed at dissecting the functional effects of LRb signaling in discrete populations of LRb neurons and the neuronal circuits that they affect will certainly improve our understanding of the LRb function in the control of eating and regulation of body weight.

References

1 Chua Jr SC, Koutras IK, Han L, Liu SM, Kay J, Young SJ, Chung WK, Leibel RL: Fine structure of the murine leptin receptor gene: splice site suppression is required to form two alternatively spliced transcripts. Genomics 1997;45:264–270.

2 Tartaglia LA: The leptin receptor. J Biol Chem 1997;272:6093–6096.

3 Kowalski TJ, Liu SM, Leibel RL, Chua Jr SC: Transgenic complementation of leptin-receptor deficiency. I. Rescue of the obesity/diabetes phenotype of LEPR-null mice expressing a LEPR-B transgene. Diabetes 2001;50:425–435.

4 Blevins JE, Baskin DG: Hypothalamic-brainstem circuits controlling eating and energy homeostasis; in Geary N, Langhans W (eds): Frontiers in Eating and Weight Regulation. Forum Nutr. Basel, Karger, 2010, vol 63, pp 133–140.

5 Langhans W, Geary N: Overview of the physiological control of eating; in Langhans W, Geary N (eds): Frontiers in Eating and Weight Regulation. Forum Nutr. Basel, Karger, 2010, vol 63, pp 9–53.

6 Myers MG, Cowley MA, Munzberg H: Mechanisms of leptin action and leptin resistance. Annu Rev Physiol 2008;70:537–556.

7 Schwartz MW, Woods SC, Porte Jr D, Seeley RJ, Baskin DG: Central nervous system control of food intake. Nature 2000;404:661–671.

8 Butler AA, Cone RD: The melanocortin receptors: lessons from knockout models. Neuropeptides 2002; 36:77–84.

9 Cowley MA, Smart JL, Rubinstein M, Cerdan MG, Diano S, Horvath TL, Cone RD, Low MJ: Leptin activates anorexigenic POMC neurons through a neural network in the arcuate nucleus. Nature 2001; 411:480–484.

10 Bouret SG, Draper SJ, Simerly RB: Trophic action of leptin on hypothalamic neurons that regulate feeding. Science 2004;304:108–110.
11 Jacobowitz DM, O'Donohue TL: Alpha-melanocyte stimulating hormone: immunohistochemical identification and mapping in neurons of rat brain. Proc Natl Acad Sci USA 1978;75:6300–6304.
12 Mountjoy KG, Mortrud MT, Low MJ, Simerly RB, Cone RD: Localization of the melanocortin-4 receptor (MC4-R) in neuroendocrine and autonomic control circuits in the brain. Mol Endocrinol 1994;8:1298–1308.
13 Elmquist JK, Bjorbaek C, Ahima RS, Flier JS, Saper CB: Distributions of leptin receptor mRNA isoforms in the rat brain. J Comp Neurol 1998;395:535–547.
14 Dhillon H, Zigman JM, Ye C, et al: Leptin directly activates SF1 neurons in the VMH, and this action by leptin is required for normal body-weight homeostasis. Neuron 2006;49:191–203.
15 Fulton S, Pissios P, Manchon RP, Stiles L, Frank L, Pothos EN, Maratos-Flier E, Flier JS: Leptin regulation of the mesoaccumbens dopamine pathway. Neuron 2006;51:811–822.
16 Hommel JD, Trinko R, Sears RM, Georgescu D, Liu ZW, Gao XB, Thurmon JJ, Marinelli M, DiLeone RJ: Leptin receptor signaling in midbrain dopamine neurons regulates feeding. Neuron 2006;51:801–810.
17 Leshan RL, Louis GW, Jo YH, Rhodes CJ, Munzberg H, Myers Jr MG: Direct innervation of GnRH neurons by metabolic- and sexual odorant-sensing leptin receptor neurons in the hypothalamic ventral premammillary nucleus. J Neurosci 2009;29:3138–3147.
18 Coppari R, Ichinose M, Lee CE, et al: The hypothalamic arcuate nucleus: a key site for mediating leptin's effects on glucose homeostasis and locomotor activity. Cell Metab 2005;1:63–72.
19 Balthasar N, Coppari R, McMinn J, et al: Leptin receptor signaling in POMC neurons is required for normal body weight homeostasis. Neuron 2004;42:983–991.
20 van de Wall E, Leshan R, Xu AW, et al: Collective and individual functions of leptin receptor modulated neurons controlling metabolism and ingestion. Endocrinology 2008;149:1773–1785.
21 Myers Jr MG: Leptin receptor signaling and the regulation of mammalian physiology. Recent Prog Horm Res 2004;59:287–304.
22 Bjorbaek C, Elmquist JK, Frantz JD, Shoelson SE, Flier JS: Identification of SOCS-3 as a potential mediator of central leptin resistance. Mol Cell 1998;1:619–625.
23 Munzberg H, Huo L, Nillni EA, Hollenberg AN, Bjorbaek C: Role of signal transducer and activator of transcription 3 in regulation of hypothalamic proopiomelanocortin gene expression by leptin. Endocrinology 2003;144:2121–2131.
24 Bjorbak C, Lavery HJ, Bates SH, Olson RK, Davis SM, Flier JS, Myers Jr MG: SOCS3 mediates feedback inhibition of the leptin receptor via Tyr985. J Biol Chem 2000;275:40649–40657.
25 Bates SH, Stearns WH, Dundon TA, et al: STAT3 signalling is required for leptin regulation of energy balance but not reproduction. Nature 2003;421:856–859.
26 Bjornholm M, Munzberg H, Leshan RL, Villanueva EC, Bates SH, Louis GW, Jones JC, Ishida-Takahashi R, Bjorbaek C, Myers Jr MG: Mice lacking inhibitory leptin receptor signals are lean with normal endocrine function. J Clin Invest 2007;117:1354–1360.
27 Cota D, Proulx K, Smith KA, Kozma SC, Thomas G, Woods SC, Seeley RJ: Hypothalamic mTOR signaling regulates food intake. Science 2006;312:927–930.
28 Minokoshi Y, Kim YB, Peroni OD, Fryer LG, Muller C, Carling D, Kahn BB: Leptin stimulates fatty-acid oxidation by activating AMP-activated protein kinase. Nature 2002;415:339–343.
29 Niswender KD, Morton GJ, Stearns WH, Rhodes CJ, Myers Jr MG, Schwartz MW: Intracellular signalling. Key enzyme in leptin-induced anorexia. Nature 2001;413:794–795.
30 Bouret SG: Development of hypothalamic neural networks for controlling appetite; in Langhans W, Geary N (eds): Frontiers in Eating and Weight Regulation. Forum Nutr. Basel, Karger, 2010, vol 63, pp 84–93.
31 Kokoeva MV, Yin H, Flier JS: 2005. Neurogenesis in the hypothalamus of adult mice: potential role in energy balance. Science 310:679–683.
32 Pinto S, Roseberry AG, Liu H, Diano S, Shanabrough M, Cai X, Friedman JM, Horvath TL: Rapid rewiring of arcuate nucleus feeding circuits by leptin. Science 2004;304:110–115.
33 Heymsfield SB, Greenberg AS, et al: Recombinant leptin for weight loss in obese and lean adults: a randomized, controlled, dose-escalation trial. JAMA 1999;282:1568–1575.
34 Considine RV, Sinha MK, Heiman ML, et al: Serum immunoreactive-leptin concentrations in normal-weight and obese humans. N Engl J Med 1996;34:292–295.
35 El Haschimi K, Pierroz DD, Hileman SM, Bjorbaek C, Flier JS: Two defects contribute to hypothalamic leptin resistance in mice with diet-induced obesity. J Clin Invest 2000;105:1827–1832.

36 Scarpace PJ, Matheny M, Moore RL, Tumer N: Impaired leptin responsiveness in aged rats. Diabetes 2000;49:431–435.

37 Tups A, Ellis C, Moar KM, Logie TJ, Adam CL, Mercer JG, Klingenspor M: Photoperiodic regulation of leptin sensitivity in the Siberian hamster, *Phodopus sungorus*, is reflected in arcuate nucleus SOCS-3 (suppressor of cytokine signaling) gene expression. Endocrinology 2004;145:1185–1193.

38 Levin BE, Dunn-Meynell AA, Banks WA: Obesity-prone rats have normal blood-brain barrier transport but defective central leptin signaling before obesity onset. Am J Physiol Regul Integr Comp Physiol 2004;286:R143–R150.

39 Banks WA, DiPalma CR, Farrell CL: Impaired transport of leptin across the blood-brain barrier in obesity. Peptides 1999;20:1341–1345.

40 Munzberg H, Flier JS, Bjorbaek C: Region-specific leptin resistance within the hypothalamus of diet-induced obese mice. Endocrinology 2004;145:4880–4889.

41 Bouret SG, Gorski JN, Patterson CM, Chen S, Levin BE, Simerly RB: Hypothalamic neural projections are permanently disrupted in diet-induced obese rats. Cell Metab 2008;7:179–185.

42 Enriori PJ, Evans AE, Sinnayah P, et al: Diet-induced obesity causes severe but reversible leptin resistance in arcuate melanocortin neurons. Cell Metab 2007;5:181–194.

43 Banks WA, Kastin AJ, Huang W, Jaspan JB, Maness LM: Leptin enters the brain by a saturable system independent of insulin. Peptides 1996;17:305–311.

44 Banks WA: Blood-brain barrier as a regulatory interface; in Langhans W, Geary N (eds): Frontiers in Eating and Weight Regulation. Forum Nutr. Basel, Karger, 2010, vol 63, pp 102–110.

45 Kievit P, Howard JK, Badman MK, Balthasar N, Coppari R, Mori H, Lee CE, Elmquist JK, Yoshimura A, Flier JS: Enhanced leptin sensitivity and improved glucose homeostasis in mice lacking suppressor of cytokine signaling-3 in POMC-expressing cells. Cell Metab 2006;4:123–132.

46 Zabolotny JM, Bence-Hanulec KK, Stricker-Krongrad A, et al: PTP1B regulates leptin signal transduction in vivo. Dev Cell 2002;2:489–495.

47 Picardi PK, CalegariVC, Prada PO, Moraes JC, Araujo E, Marcondes MC, Ueno M, Carvalheira JB, Velloso LA, Saad MJ: Reduction of hypothalamic protein tyrosine phosphatase improves insulin and leptin resistance in diet-induced obese rats. Endocrinology 2008;149:3870–3880.

48 White CL, Whittington A, Barnes MJ, Wang Z, Bray GA, Morrison CD: HF diets increase hypothalamic PTP1B and induce leptin resistance through both leptin-dependent and -independent mechanisms. Am J Physiol Endocrinol Metab 2009;296:E291–E299.

49 Wellen KE, Hotamisligil GS: Obesity-induced inflammatory changes in adipose tissue. J Clin Invest 2003;112:1785–1788.

50 Ogus S, Ke Y, Qiu J, Wang B, Chehab FF: Hyperleptinemia precipitates diet-induced obesity in transgenic mice overexpressing leptin. Endocrinology 2003;144:2865–2869.

51 Sahu A: Resistance to the satiety action of leptin following chronic central leptin infusion is associated with the development of leptin resistance in neuropeptide Y neurones. J Neuroendocrinol 2002;14:796–804.

52 Scarpace PJ, Tumer N: Peripheral and hypothalamic leptin resistance with age-related obesity. Physiol Behav 2001;74:721–727.

53 Rahmouni K, Haynes WG, Morgan DA, Mark AL: Selective resistance to central neural administration of leptin in agouti obese mice. Hypertension 2002;39:486–490.

54 Faouzi M, Leshan R, Bjornholm M, Hennessey T, Jones J, Munzberg H: Differential accessibility of circulating leptin to individual hypothalamic sites. Endocrinology 2007;148:5414–5423.

55 Ladyman SR, Grattan DR: Region-specific reduction in leptin-induced phosphorylation of signal transducer and activator of transcription-3 (STAT3) in the rat hypothalamus is associated with leptin resistance during pregnancy. Endocrinology 2004;145:3704–3711.

56 Diano S, Kalra SP, Sakamoto H, Horvath TL: Leptin receptors in estrogen receptor-containing neurons of the female rat hypothalamus. Brain Res 1998;812:256–259.

Heike Münzberg, PhD
Pennington Biomedical Research Center, LSU System
6400 Perkins Rd
Baton Rouge, LA 70808 (USA)
Tel. +1 225 763 2769, Fax +1 225 763 0260, E-Mail heike.munzberg@pbrc.edu

Langhans W, Geary N (eds): Frontiers in Eating and Weight Regulation.
Forum Nutr. Basel, Karger, 2010, vol 63, pp 133–140

Hypothalamic-Brainstem Circuits Controlling Eating

James E. Blevins · Denis G. Baskin

Research and Development Service, Department of Veterans Affairs Puget Sound Health Care System, and Department of Medicine, Division of Metabolism, Endocrinology, and Nutrition, University of Washington, Seattle, Wash., USA

Abstract

It is now axiomatic that neurons in the hypothalamic arcuate nucleus have a primary role in responding to changes in circulating levels of leptin and transmitting signals to downstream circuits that influence eating and energy expenditure. Signals generated from the gastrointestinal tract during meals reach the brainstem, via the vagus nerve and other routes, and impinge on neural circuits that influence the timing and size of meals and amount of food consumed. One of the mechanisms by which leptin exerts its anorexic effects is by increasing the effectiveness of intestinal signals that cause satiation during a meal. It is clear that the effects of gut satiation signals such as CCK can be amplified by leptin acting in the CNS, and in the arcuate nucleus in particular. The present article describes the state of our knowledge about specific neural circuits between the hypothalamus and brainstem that play a role in the interaction of leptin and meal-control signals to control food intake.

Many years of research have established that energy homeostasis is regulated by the central nervous system (CNS) through the control of eating and energy expenditure. This process involves the integration of very different types of sensory information. One type, 'meal-control signals', involves afferent information arising in relation to eating that affect meal timing and size. A prototypical example is cholecystokinin (CCK), a gut peptide that is released from the intestines during meals and acts to produce meal-ending satiation. The CCK satiation signal and many other meal-control signals reach the nucleus tractus solitarius (NTS) in the caudal brainstem, via the vagus nerve and other routes. In contrast to meal-control signals, 'adiposity signals' are humoral signals generated in proportion to adipose tissue mass that directly affect CNS neurons and circuits that regulate energy homeostasis. The best researched examples of these are leptin and insulin. Each of these hormones is thought to act via receptors in the hypothalamic arcuate nucleus (Arc) to affect eating and energy expenditure over the longer term. The identity of the neurons in the CNS circuits processing meal-control and adiposity signaling has been the subject of intense research in recent years, and the literature in this area has been reviewed in this volume [1, 2] and elsewhere [3, 4].

Clearly, for adiposity signals to control eating, they must interact in the CNS with the neural representations of meal-control signals. It is now axiomatic, first, that Arc neurons have a primary role in responding to changes in circulating levels of leptin, insulin, and other metabolic and endocrine adiposity signals related to energy homeostasis and transmitting signals to downstream circuits that influence eating and energy expenditure and, second, that one of the mechanisms by which adiposity signals exert their anorexic effects is by increasing the effectiveness of gastrointestinal meal-control signals that cause satiation or satiety. Only recently, however, have details about the neural circuits mediating this interaction begun to emerge [3–6]. Therefore, in the present chapter we describe the state of our knowledge of the specific hypothalamic-brainstem neural circuits that mediate the interaction between meal-control and adiposity signals to control eating and energy homeostasis. We focus on CCK and leptin, the signals for which the most information regarding neural processing is available.

Functional Interactions of Adiposity and Satiation Signals

Several lines of evidence indicate that adiposity signals increase the potency of CCK and other meal-control signals. Exogenous leptin increased the ability of gastric loads or exogenous CCK-8 to inhibit eating [5, 6] and to enhance brainstem neuronal activation, as measured by c-Fos expression [5–8]. Leptin also enhanced the food intake and brainstem c-Fos responses to bombesin [9] (an anurian homolog to the mammalian brain-gut peptides gastrin-releasing peptide and neuromedin B), as well as the eating responses to PYY(3–36) [10] and GLP-1 [11]. Likewise, in transgenic rats with impaired leptin signaling, restoration of leptin signaling specifically to the Arc was sufficient to enhance the satiating potency and the c-Fos response to CCK-8 [12]. A similar decrease in CCK-8's satiating potency as in leptin receptor-deficient rats occurred in fasting rats, in which endogenous leptin levels are very low [13].

Other data indicate that hypothalamic-brainstem connections may be sufficient, but are not necessary, for CCK to inhibit eating. First, the eating-inhibitory effect of CCK-8 in neonatal rats was accompanied by c-Fos induction in the brainstem, but not the hypothalamus, suggesting that a neural connection between the forebrain and brainstem is not essential in order to respond to CCK [14]. Second, decerebrate rats, in which all neural connections between hypothalamus and brainstem are severed, also had intact eating responses to CCK and other short-term meal-related stimuli, but they are not able to mount a normal compensatory response to fasting-induced energy deficits by increasing food intake [15].

Neural Connections between the Hypothalamus and NTS

Descending Arc-NTS Projections

How does hypothalamic leptin signaling communicate with brainstem NTS neurons that respond to satiation signals, such as CCK? We address this question with a discussion of some of the known descending forebrain-brainstem neuronal circuits likely involved with transmitting hypothalamic leptin signaling to brainstem neurons sensitive to satiation signals. Studies have shown that α–melanocyte-stimulating hormone (α-MSH) projections originating from proopiomelanocortin (POMC) neurons in the Arc terminate in close anatomical proximity to neurons in the NTS that are sensitive to gastric distension [16], confirming that α-MSH projections from the Arc to the NTS exist. Moreover, endogenous melanocortin signaling in the brainstem contributes to the eating-inhibitory response to CCK-8 [17]. Together, these findings raise the possibility that melanocortin input from the Arc-NTS pathway containing α-MSH is an important mechanism to explain how leptin signaling in the hypothalamus communicates with key neurons in the brainstem sensitive to CCK.

Descending Paraventricular Nucleus-NTS Projections

Several findings support a role for oxytocin and corticotrophin-releasing factor (CRF) neurons in the paraventricular nucleus (PVN) in the control of eating and regulation of body weight. These peptides each inhibit eating when injected intracerebroventricularly (ICV) [18, 19]. Oxytocin gene expression is decreased in *Sim1* haploinsufficient mice, a condition accompanied by hyperphagia and obesity [20]. Both oxytocin and CRF peptides induce c-Fos in brainstem areas involved in controlling eating and receptors for both peptides are found in the NTS [21, 22].

Leptin-induced activation of melanocortin-sensing circuits in the PVN may explain how hypothalamic leptin signaling communicates with PVN neurons that project to brainstem neurons sensitive to CCK and other meal-control signals. PVN neurons express melanocortin receptors (MC3R and MC4R) and the PVN is an important site for the action of melanocortins on eating [23]. Subsets of neurons in the PVN that project to the NTS express MC4R mRNA [24] and/or oxytocin, and these oxytocin neurons increase c-Fos expression in response to leptin [7]. Oxytocin fibers in the medial NTS are distributed in close anatomical proximity to neurons that are activated to express c-Fos by exogenous administration of CCK-8 [25]. Central administration of an oxytocin receptor antagonist attenuates the ability of leptin to enhance this neuronal response to CCK-8 in the medial NTS [7]. Oxytocin fibers in the NTS also are found in close anatomical proximity to GLP-1 neurons in the NTS [26]. Furthermore, ICV administration of oxytocin induces c-Fos expression in GLP-1 neurons in the NTS, whereas ICV administration of a GLP-1R receptor antagonist blocks the effects of ICV oxytocin to inhibit eating [27], a finding that supports the existence of a PVN-NTS oxytocin projection linking hypothalamic leptin signaling with brainstem neurons that control food intake. Taken together, these findings raise the possibility that a leptin-sensitive melanocortin projection to the PVN activates oxytocin neurons that project to brainstem neurons that are sensitive to meal-control signals.

Leptin may also influence brainstem satiation circuitries through PVN CRF neurons [28]. MC3R/MC4Rs are found on CRF neurons in the PVN [29], and PVN CRF neurons have direct projections to the NTS [30, 31]. Leptin activation of STAT3 phosphorylation, a marker of direct leptin action, has also been shown to increase thyroid releasing hormone (TRH) mRNA levels in a subpopulation of PVN neurons [32], consistent with a putative role for TRH in the control of eating and energy metabolism [33]. It is not known, however, whether the leptin responsive TRH neurons in the PVN contribute to the TRH present in the NTS [34]. Finally although leptin activates c-Fos expression in a subpopulation of oxytocin PVN neurons that project to the NTS [7] and immunocytochemical evidence for leptin receptor expression in the PVN has been reported [35], it remains to be demonstrated whether PVN oxytocin neurons are activated directly by leptin or indirectly downstream of primary leptin action in the Arc or elsewhere.

Descending Lateral Hypothalamus-NTS Projections

Leptin-sensitive POMC and NPY neurons in the Arc innervate melanin-concentrating hormone (MCH) and orexin neurons in the lateral hypothalamus (LH) [36]. Descending projections from the LH to the NTS contain orexin [37] and MCH, and these MCH fibers innervate neurons in the NTS activated by gastric nutrient loads [38]. Thus, it is possible that leptin signaling to the Arc could be relayed to NTS neurons that are sensitive to meal-control signals via MCH and orexin neurons in the LH. Leptin may also directly inhibit descending LH-NTS projections, as LH leptin administration decreases food intake [39] and leptin receptors are expressed by MCH and orexin neurons [40, 41].

Descending Arc-Parabrachial Nucleus Projections

Recent studies have demonstrated a critical role in the control of food intake for a projection containing agouti-related peptide (AGRP)/gamma-amino butyric acid (GABA) from the Arc to the parabrachial nucleus (PBN) [42]. Lesions of the lateral PBN attenuate conditioned taste aversions [43], as well as the eating-inhibitory effects of amylin and CCK-8 [44]. Whether these lesions alter CCK-8-elicited satiation through interruptions of ascending CCK-8-ergic projections from the PBN to the ventromedial hypothalamus [45] or whether they impair the input of descending Arc-PBN pathways remain to be determined.

Ascending NTS-PVN Projections

Ascending NTS-PVN projections containing norepinephrine [46], neuropeptide Y (NPY) [47], and GLP-1 [48] are implicated in the control of eating. Intra-PVN administration of the cellular toxin saporin that is conjugated to a monoclonal antibody against dopamine β-hydroxylase destroys catecholamine (norepinephrine/epinephrine) neuronal projections from brainstem catecholamine neurons to the PVN, including those in the A1/C1 of the ventrolateral medulla and in the A2/C2 region of the caudal medial NTS [49]. NPY is coexpressed in a subset of these catecholaminergic projections to the PVN, including A1/C1 and C2 [47].

CCK-8 activates catecholamine neurons and NPY neurons in the NTS [50], and many of the CCK-8-activated catecholamine neurons project from the A2 cell group in the NTS to the PVN and amygdala [51]. Selective lesions of the catecholamine neurons in the NTS result in decreased c-Fos induction in oxytocin neurons in the PVN and an attenuated eating-inhibitory response to CCK-8 but did not alter c-Fos expression in the PBN or the central nucleus of the amygdala [52]. Thus, while reciprocal catecholamine connections appear to exist between the hypothalamus and NTS, the circuitry is not well understood and other connections are likely to be important in the control of eating. For example, CCK-8 activates PVN oxytocin neurons that project back to the NTS and nearby nuclei [53]. In addition, CCK-8 activates GLP-1 neurons in the NTS with ascending projections to the PVN [48]. These findings indicate that activation of ascending noradrenergic (and possibly GLP-1) projections from the NTS and subsequent activation of PVN oxytocin neurons that project to the NTS are associated with the satiating effect of CCK-8.

Earlier findings indicate that leptin (or gastric load) in combination with CCK-8 result in more c-Fos expression in the PVN compared to either treatment alone [6, 8, 9]. Another potential mechanism to explain how leptin enhances CCK-8-elicited satiation may involve dual activation of oxytocin neurons in the PVN by, first, activation of CCK-sensitive ascending NTS-PVN noradrenergic projections onto PVN oxytocin neurons containing alpha1 adrenergic receptors [54] and, second, leptin-induced activation of Arc projections to PVN oxytocin neurons [2–4, 7]; a population of these dually activated oxytocin neurons projects to the NTS to control eating. Indeed, leptin administration enhances CCK-8-elicited c-Fos induction and c-Fos mRNA expression in tyrosine hydroxylase-immunoreactive catecholamine neurons in the A2/C2 region of the caudal medial NTS [8]. Thus, evidence indicates that a subpopulation of CCK-sensitive catecholamine neurons in the A2/C2 region of the NTS with ascending projections to the PVN (either directly from the NTS or indirectly from the ventrolateral medulla), may contribute to the ability of leptin to enhance the satiety response to CCK-8.

Conclusion

The control of eating and regulation of body weight involves complex CNS neuronal circuits. This review has focused mainly on the controls exerted by the adiposity signal leptin and the meal-control signal CCK. Intact reciprocal neuronal connections between brainstem (especially NTS, PBN) and hypothalamic (especially Arc, PVN and LH) areas are required for the coordinate control of eating by leptin, and perhaps other adiposity signals, and CCK, and perhaps other meal-control signals. This view of the link between homeostatic and meal-control signals is becoming more complicated, however, as new information emerges concerning, for example, the potential role of leptin signaling to ventral telencephalic reward systems. The mechanisms that integrate neuronal crosstalk among the hypothalamus, brainstem, telencephalic reward systems and other brain regions provide a fertile ground for future research.

Acknowledgements

The preparation of this article is supported by facilities at the VA Puget Sound Health Care System, Seattle, Wash., USA. Dr. Baskin is a VA Senior Research Career Scientist, and Dr. Blevins is a VA Research Biologist, in the Research and Development Service, Department of Veterans Affairs Puget Sound Health Care System, Seattle, Wash., USA. This research was supported by research funds from the Veterans Health Administration of the US Department of Veterans Affairs and NIH NIDDK grant P30 DK-17047 through the Cellular and Molecular Imaging Core of the University of Washington Diabetes Endocrinology Research Center.

References

1 Langhans W, Geary N: Overview of the physiological control of eating; in Langhans W, Geary N (eds): Frontiers in Eating and Weight Regulation. Forum Nutr. Basel, Karger, 2010, vol 63, pp 9–53.

2 Schwartz GJ: Brainstem integrative function in the central nervous system control of food intake; in Langhans W, Geary N (eds): Frontiers in Eating and Weight Regulation. Forum Nutr. Basel, Karger, 2010, vol 63, pp 141–151.

3 Morton GJ, Cummings DE, Baskin DG, Barsh GS, Schwartz MW: Central nervous system control of food intake and body weight. Nature 2006;443:289–295.

4 Schwartz MW, Woods SC, Porte D Jr, Seeley RJ, Baskin DG: Central nervous system control of food intake. Nature 2000;404:661–67.

5 Emond M, Schwartz GJ, Ladenheim EE, Moran TH: Central leptin modulates behavioral and neural responsivity to CCK. Am J Physiol Regul Integr Comp Physiol 1999;276:R1545–1549.

6 Emond M, Ladenheim EE, Schwartz GJ, Moran TH: Leptin amplifies the feeding inhibition and neural activation arising from a gastric nutrient preload. Physiol Behav 2001;72:123–128.

7 Blevins JE, Schwartz MW, Baskin DG: Evidence that paraventricular nucleus oxytocin neurons link hypothalamic leptin action to caudal brain stem nuclei controlling meal size. Am J Physiol Regul Integr Comp Physiol 2004;287:R87–R96.

8 Williams DL, Schwartz MW, Bastian LS, Blevins JE, Baskin DG: Immunocytochemistry and laser capture microdissection for real-time quantitative PCR identify brainstem neurons activated by interaction between leptin and cholecystokinin. J Histochem Cytochem 2008;56:285–293.

9 Ladenheim EE, Emond, E, Moran TH: Leptin enhances feeding suppression and neural activation produced by systemically administered bombesin. Am J Physiol Regul Integr Comp Physiol 2005;89: R473–R477.

10 Unniappan S, Kieffer TJ: Leptin extends the anorectic effects of chronic PYY(3–36) administration in ad libitum-fed rats. Am J Physiol Regul Integr Comp Physiol 2008;295:R51–R58.
11 Williams DL, Baskin DG, Schwartz MW: Leptin regulation of the anorexic response to glucagon-like peptide-1 receptor stimulation. Diabetes 2006;55:3387–3393.
12 Morton GJ, Blevins JE, Williams DL, Niswender KD, Gelling RW, Rhodes CJ, Baskin DG, Schwartz MW: Leptin action in the forebrain regulates the brainstem response to satiety signals. J Clin Invest 2005;115:703–10.
13 McMinn JE, Sindelar DK, Havel PJ, Schwartz MW: Leptin deficiency induced by fasting impairs the satiety response to cholecystokinin. Endocrinology 2000;141:4442–4448.
14 Rinaman L, Hoffman GE, Stricker EM, Verbalis JG: Exogenous cholecystokinin activates cFos expression in medullary but not hypothalamic neurons in neonatal rats. Dev Brain Res 1994;77:140–145.
15 Kaplan JM, Seeley RJ, Grill HJ: Daily caloric intake in intact and chronic decerabrate rats. Behav Neurosci 1993;107:876–881.
16 Zheng H, Patterson LM, Pfifer CB, Berthoud, H-R: Brainstem melanocortinergic modulation of meal size and identification of hypothalamic POMC projections. Am J Physiol Regul Integr Comp Physiol 2005;289:R247–R258.
17 Fan W, Ellacott KL, Halatchev IG, Takahashi K, Yu P, Cone RD: Cholecystokinin-mediated suppression of feeding involves the brainstem melanocortin system. Nat Neurosci 2004;7:335–336.
18 Arletti R, Benelli A, Bertolini A: Influence of oxytocin on feeding behavior in the rat. Peptides 1989;10:89–93.
19 Gosnell BA, Morley JE, Levine AS: A comparison of the effects of corticotropin releasing factor and sauvagine on food intake. Pharmacol Biochem Behav 1983;19:771–775.
20 Kublaoui BM, Gemelli T, Tolson KP, Wang Y, Zinn AR: Oxytocin deficiency mediates hyperphagic obesity of Sim1 haploinsufficient mice. Mol Endocrinol 2008;22:1723–1734.
21 Benoit SC, Thiele TE, Heinrichs SC, Rushing PA, Blake KA, Seeley RJ: Comparison of central administration of corticotropin-releasing hormone and urocortin on food intake, conditioned taste aversion, and c-Fos expression. Peptides 2000;21:345–351.
22 Olson BR, Freilino M, Hoffman GE, Stricker EM, Sved AF, Verbalis JG: c-Fos expression in rat brain and brainstem nuclei in response to treatments that alter food intake and gastric motility. Mol Cell Neurosci 1993;4:93–106.
23 Seeley RJ, Yagaloff KA, Fisher SL, Burn P, Thiele TE, van Dijk G, Baskin DG, Schwartz MW: Melanocortin receptors in leptin effects. Nature 1997;390:349.
24 Blevins JE, Morton GJ, Williams DL, Caldwell DW, Bastian LS, Wisse BE, Schwartz MW, Baskin DG: Forebrain melanocortin signaling enhances the brainstem satiety response to CCK-8. Am J Physiol Regul Integr Comp Physiol 2009;296:R476–R484.
25 Blevins JE, Eakin TJ, Murphy JA, Schwartz MW, Baskin DG: Oxytocin innervation of caudal brainstem nuclei activated by cholecystokinin. Brain Res 2003;993:30–41.
26 Rinaman L, Rothe EE: GLP-1 receptor signaling contributes to anorexigenic effect of centrally administered oxytocin in rats. Am J Physiol Regul Integr Comp Physiol 2002;283:R99–R106.
27 Goldstone AP, Mercer JG, Gunn I, Moar KM, Edwards CM, Rossi M, Howard JK, Rasheed S, Turton MD, Small C, Heath MM, O'Shea D, Steere J, Meeran K, Ghatei MA, Hoggard N, Bloom SR: Leptin interacts with glucagon-like peptide-1 neurons to reduce food intake and body weight in rodents. FEBS Lett 1997;415:134–138.
28 Schwartz MW, Seeley RJ, Campfield LA, Burn P, Baskin DG: Identification of targets of leptin action in rat hypothalamus. J Clin Invest 1996;98:1101–1106.
29 Lu XY, Barsh GS, Akil H, Watson SJ: Interaction between alpha-melanocyte-stimulating hormone and corticotropin-releasing hormone in the regulation of feeding and hypothalamo-pituitary-adrenal responses. J Neurosci 2003;23:7863–7872.
30 Sawchenko PE, Swanson LW: Immunohistochemical identification of neurons in the paraventricular nucleus of the hypothalamus that project to the medulla or to the spinal cord in the rat. J Comp Neurol 1982;205:260–272.
31 Sawchenko PE, Swanson LW: Localization, colocalization, and plasticity of corticotropin-releasing factor immunoreactivity in rat brain. Fed Proc 1985;44:221–227.
32 Huo L, Münzberg H, Nillni EA, Bjørbaek C: Role of signal transducer and activator of transcription 3 in regulation of hypothalamic *trh* gene expression by leptin. Endocrinology 2004;145:2516–2523.
33 Lechan RM, Fekete C: The TRH neuron: a hypothalamic integrator of energy metabolism. Prog Brain Res 2006;153:209–235.
34 Siaud P, Tapia-Arancibia L, Szafarczyk A, Alonso G: Increase of thyrotropin-releasing hormone immunoreactivity in the nucleus of the solitary tract following bilateral lesions of the hypothalamic paraventricular nuclei. Neurosci Lett 1987;79:47–52.
35 Hakansson ML, Brown H, Ghilardi N, Skoda RC, Meister B: Leptin receptor immunoreactivity in chemically defined target neurons of the hypothalamus. J Neurosci 1998;18:559–572.

36 Broberger C, De Lecea L, Sutcliffe JG, Hökfelt T: Hypocretin/orexin- and melanin-concentrating hormone-expressing cells form distinct populations in the rodent lateral hypothalamus: relationship to the neuropeptide Y and agouti gene-related protein systems. J Comp Neurol 1998;402:460–474.
37 Peyron C, Tighe DK, van den Pol AN, de Lecea L, Heller HC, Sutcliffe JG, Kilduff TS: Neurons containing hypocretin (orexin) project to multiple neuronal systems. J Neurosci 1998;18:9996–10015.
38 Zheng H, Patterson LM, Morrison C, Banfield BW, Randall JA, Browning KN, Travagli RA, Berthoud HR: Melanin concentrating hormone innervation of caudal brainstem areas involved in gastrointestinal functions and energy balance. Neuroscience 2005;135:611–625.
39 Satoh N, Ogawa Y, Katsuura G, Hayase M, Tsuji T, Imagawa K, Yoshimasa Y, Nishi S, Hosoda K, Nakao K: The arcuate nucleus as a primary site of satiety effect of leptin in rats. Neurosci Lett 1997;224:149–152.
40 Iqbal J, Pompolo S, Murakami T, Grouzmann E, Sakurai T, Meister B, Clarke IJ: Immunohistochemical characterization of localization of long-form leptin receptor (OB-Rb) in neurochemically defined cells in the ovine hypothalamus. Brain Res 2001;920:55–64.
41 Hakansson M, de Lecea L, Sutcliffe JG, Yanagisawa M, Meister B: Leptin receptor- and STAT3-immunoreactivities in hypocretin/orexin neurones of the lateral hypothalamus. J Neuroendocrinol 1999;11:653–663.
42 Wu Q, Boyle MP, Palmiter RD: Loss of GABAergic signaling by AgRP neurons to the parabrachial nucleus leads to starvation. Cell 2009;137:1225–1234.
43 Reilly S: The parabrachial nucleus and conditioned taste aversion. Brain Res Bull 1999;48:239–254.
44 Becskei C, Grabler V, Edwards GL, Riediger T, Lutz TA: Lesion of the lateral parabrachial nucleus attenuates the anorectic effect of peripheral amylin and CCK. Brain Res 2007;1162:76–84.
45 Fulwiler CE, Saper CB: Cholecystokinin-immunoreactive innervation of the ventromedial hypothalamus in the rat: possible substrate for autonomic regulation of feeding. Neurosci Lett 1985; 53:289–296.
46 Cunningham ET Jr, Sawchenko PE: Anatomical specificity of noradrenergic inputs to the paraventricular and supraoptic nuclei of the rat hypothalamus. J Comp Neurol 1988;274:60–76.
47 Sawchenko PE, Swanson LW, Grzanna R, Howe PR, Bloom SR, Polak JM: Colocalization of neuropeptide Y immunoreactivity in brainstem catecholaminergic neurons that project to the paraventricular nucleus of the hypothalamus. J Comp Neurol 1985;241:138–153.
48 Rinaman L: Interoceptive stress activates glucagon-like peptide-1 neurons that project to the hypothalamus. Am J Physiol Regul Integr Comp Physiol 1999;277:R582–R590.
49 Estacio MA, Tsukamura H, Reyes BA, Uenoyama Y, I'anson H, Maeda K: Involvement of brainstem catecholaminergic inputs to the hypothalamic paraventricular nucleus in estrogen receptor alpha expression in this nucleus during different stress conditions in female rats. Endocrinology 2004;145: 4917–4926.
50 Rinaman L, Verbalis JG, Stricker EM, Hoffman GE: Distribution and neurochemical phenotypes of caudal medullary neurons activated to express cFos following peripheral administration of cholecystokinin. J Comp Neurol 1993;338:475–490.
51 Rinaman L, Hoffman GE, Dohanics J, Le WW, Stricker EM, Verbalis JG: Cholecystokinin activates catecholaminergic neurons in the caudal medulla that innervate the paraventricular nucleus of the hypothalamus in rats. J Comp Neurol 1995;360:246–256.
52 Rinaman L: Brainstem noradrenergic lesions attenuate anorexia and alter central cfos expression in rats after gastric viscerosensory stimulation. J Neurosci 2003;23:10084–10092.
53 Olson BR, Hoffman GE, Sved AF, Stricker EM, Verbalis JG: Cholecystokinin induces c-fos expression in hypothalamic oxytocinergic neurons projecting to the dorsal vagal complex. Brain Res 1992; 569:238–248.
54 Wellman PJ, Davies BT, Morien A, McMahon L: Modulation of feeding by hypothalamic paraventricular nucleus alpha1- and alpha2-adrenergic receptors. Life Sci 1993;53:669–679.

Denis G. Baskin, PhD
Research and Development Service, Mail Stop S-151,
Department of Veterans Affairs Puget Sound Health Care System
1660 South Columbian Way
Seattle, WA 98108 (USA)
Tel. +1 206 768 5222, +1 206 764 2138, Fax +1 206 764 2164, E-Mail baskindg@uw.edu

Langhans W, Geary N (eds): Frontiers in Eating and Weight Regulation.
Forum Nutr. Basel, Karger, 2010, vol 63, pp 141–151

Brainstem Integrative Function in the Central Nervous System Control of Food Intake

Gary J. Schwartz

Departments of Medicine and Neuroscience, Albert Einstein College of Medicine of Yeshiva University, Bronx, N.Y., USA

Abstract

The caudal brainstem lies at a critical nexus in the neural hierarchy that helps determine the negative feedback control of ingestion. It is the first central nervous system site that receives neural input from vagal and nonvagal visceral afferents that convey not only meal-related signals from postoral sites reflecting chemical, mechanical and nutrient properties of ingested foods, but also responses to nutrient stimulated peptides and neurotransmitters via extrinsic gut afferent receptors. In addition, the circumventricular area postrema (AP) affords the caudal brainstem access to circulating factors that are released during a meal, as well as to adiposity hormones that reflect the availability of stored nutrients. Brainstem neurons themselves express eating modulatory neuropeptides as well as their cognate receptors, raising the possibility that local ligand-receptor interactions contribute to the neural basis of eating behavior. Finally, forebrain hypothalamic projections extend to brainstem neurons that also respond to humoral and meal-related post-oral signals from peripheral gut afferents, providing critical descending influences in the negative feedback control of food intake [1]. This article characterizes recent advances in our understanding of how peripheral, brainstem-intrinsic and descending forebrain influences may converge in the caudal brainstem to reduce food intake, with a focus on their roles in the control of meal size.

Brainstem Representation of Gut Vagal Afferent Meal-Related Signals

Coordinated application of molecular biological, pharmacological and genetic approaches has revealed important roles for gut neuropeptides and neurotransmitters acting at gut afferent and brainstem receptors in the negative feedback control of meal size. Four neurochemical mediators of gut vagal afferent signaling are particularly relevant for this control: the serotonin-3 receptor (5HT3R), the glutamate NMDA receptor (NMDAR) and its subunits, the glucagon like peptide-1 receptor (GLP-1R), and LRb, the functional long form of the leptin receptor. This section will briefly characterize current understanding of each of these receptor subtypes at the vagal afferent/brainstem interface and their role in the control of meal size.

Serotonin 5HT3 Receptors

Serotonin is released by intestinal enterochromaffin cells upon duodenal nutrient exposure. Proximal gut vagal afferents express 5HT3R [2], 5HT rapidly activates gut vagal afferents via 5HT3R, and gut-recipient nodose ganglion neuronal responses to intestinal nutrients are blocked by 5HT3R antagonists [3]. Taken together, these data support a role for local 5HT release acting at 5HT3R in the negative feedback control of meal size. Consistent with this suggestion, peripheral administration of the prototypical 5HT3R antagonist ondansetron blocked the ability of duodenal nutrient infusions to inhibit eating and gastric emptying [4, 5], and attenuated the eating inhibitory effects of the gut satiation peptide cholecystokinin (CCK) [6].

However, as peripherally administered ondansetron readily crosses the blood brain barrier, the above results raise the possibility that central 5HT3 R also contribute to the serotonergic postoral control of meal size. The dorsal vagal complex (DVC), including the AP, the nucleus of the solitary tract (NTS), and the dorsal motor nucleus of the vagus (DMX), contains the highest density of central 5HT3R. Subdiaphragmatic vagotomy attenuates brainstem 5HT3R binding, suggesting that vagal afferent terminations in the DVC express functional 5HT3R. Fourth ventricular (4V) and discrete unilateral parenchymal administration of ondansetron directly into the medial NTS blocked the eating inhibitory actions of peripherally administered CCK [7], supporting a role for caudal brainstem 5HT3R in the postoral negative feedback control of meal size. The relative contributions of central vagal afferent and NTS neuronal 5HT3R in this control remain unknown. This gap in understanding could be addressed by assessing the eating inhibitory effects of postoral negative feedback stimuli in response to unilateral NTS 5TH3R injections, following ipsilateral vagal afferent rhizotomies, where the central transport of 5HT3R from the nodose to brainstem vagal afferent terminations would be blocked.

Glutamatergic NMDA Receptors

Nodose ganglion neurons responsive to intestinal nutrients not only express 5HT3R, but are immunopositive for glutamate [8], and express multiple subunits of functional glutamatergic NMDA receptors (NMDAR) [9, 10]. NMDA agonists directly stimulate gut vagal afferents, and potentiate their excitatory responses to mechanical gut stimuli [11]. These data suggest that glutamatergic transmission along the gut-brain axis contributes to the negative feedback control of meal size. Accordingly, peripheral administration of the NMDAR antagonist MK-801 selectively increased food but not water intake in rats and this effect required the subdiaphragmatic vagus, capsaicin-sensitive vagal afferents, and the DVC [12]. Peripheral MK-801 administration also blocked CCK satiation, supporting a role for peripheral NMDA receptors in the postoral controls of meal size [13].

However, gut nutrient- and distension-sensitive NTS neurons, the targets of centrally projecting gut vagal afferents, also express functional NMDAR [14]. This raises the possibility that local glutamatergic transmission at the brainstem vagal afferent/NTS interface mediates the control of meal size. Indeed, caudal brainstem injections of NMDAR antagonists into either the 4V or the mNTS increase food intake within the setting of an individual meal, and over 24 h [15]. The effects of brainstem NMDAR blockade on meal size require central vagal afferent terminals, as removal of the nodose ganglion ipsilateral to a NTS MK-801 injection blocks MK-801 induced eating [16]. More recent work of Covasa and colleagues have identified specific roles for glutamate receptor subunits NR2B and NR2C in the brainstem NMDAR mediation of the post-oral control of meal size. Fourth ventricular and parenchymal NTS injections of potent and selective NR2B and 2C antagonists recapitulated the effects of similarly placed MK-801 injections by increasing sucrose intake during a meal, and by blocking CCK satiety [17, 18].

The local nutrient environment, specifically extracellular glucose concentration, also helps determine the efficacy of glutamatergic and 5HT3R signaling at the brainstem vagal afferent-NTS interface. Glucose dose-dependently modulates presynaptic glutamate release from vagal afferents onto postsynaptic NTS neurons [19], and this glucose-stimulated release is blocked presynaptically by 5HT3R antagonists [20]. In addition, local glucose levels appear to influence the tonic levels of endogeonus 5HT3R-mediated NTS activity. Given the proximity of the vagal afferent-NTS interface to the AP and its relatively porous blood brain barrier, these findings suggest the intriguing possibility that local glucose concentrations increase the gain of gut afferent glutamatergic and serotonergic signals generated during a meal. The resulting increase in vagal afferent-NTS activity would provide a brainstem neural representation of increased local nutrient availability (e.g. increased glucose levels) and could thus be important to the negative feedback control of ingestion.

Glucagon-Like Peptide 1 Receptors

GLP-1 is released from intestinal enteroendocrine cells by the presence of intestinal nutrients, especially carbohydrates and fat. Peripheral administration of GLP-1 improves glucose-dependent insulin secretion and glucose homeostasis, decreases gastric emptying and reduces food intake [21]. Nodose ganglion neurons express GLP-1 receptors (GLP-1R) mRNA, and gastric and hepatic portal vagal afferents are rapidly stimulated by intravenous administration of GLP-1 at physiological levels [22]. Gastric, but not hepatic branch vagal afferent responses to GLP-1 are blocked by administration of the GLP-1R antagonist exendin (9–39) amide (Ex-9) [22, 23], suggesting a selective role for vagal afferent GLP-1R in the mediation of gastrointestinal signals important in the negative feedback control of meal size. Consistent with this possibility, subdiaphragmatic vagal deafferentation blocked the effect of intraperitoneal GLP-1 administration on meal size [24]. Taken together, these data support a role for peripheral vagal afferent GLP-1R in the negative feedback control of ingestion.

GLP-1 crosses the blood brain barrier by simple diffusion, and specific GLP-1 binding and in situ hybridization for GLP-1R within the rat central nervous system is particularly dense in the brainstem AP, NTS, and DMX [25]. Among these populations, GLP-1R are expressed by gut-recipient NTS neurons activated by multiple meal-related stimuli, including gastric distension and duodenal nutrients [26]. Peripheral administration of GLP-1 at eating inhibitory doses increases neuronal activation signaling intensity in multiple GLP-1R target areas, including the AP and the NTS, as measured by manganese enhanced magnetic resonance imaging [27]. Intracerebroventricular administration of GLP-1 selectively reduced meal size, and the brainstem is likely a critical substrate for this effect; both peripheral and brainstem 4V injections of the GLP-1 agonist Ex-4 decreased eating in chronically decerebrated rats [28]. In neurologically intact rats, 4V and parenchymal medial NTS application of Ex-9 increased food intake, and blocked the ability of gastric balloon distension to limit subsequent eating during a scheduled meal [29]. Thus, the caudal brainstem integrates gut peptide and meal-related gut vagal afferent signals to limit meal size. As GLP-1R are produced by nodose ganglion neurons and transported along vagal afferents, brainstem GLP-1 may act at central vagal afferent terminations expressing GLP-1R receptors to control meal size. For example, it is not known whether GLP-1R on central terminations of vagal afferents are necessary for the ability of brainstem Ex-9 injections to stimulate food intake, as GLP-1R exist at both pre-synaptic (gut vagal afferent) and postsynaptic (NTS) sides of the vagus/NTS interface. Unliateral brainstem GLP-1 injections following ipsilateral vagal afferent rhizotomy could help identify the relative contributions of central vagal afferent and NTS GLP-1 receptor expression to the inhibition of food intake produced by brainstem GLP-1, as well as reveal the degree to which central vs. peripheral vagal afferent GLP-1 R are important for GLP-1 inhibition of meal size.

LRb

Peripheral and central administration of the adiposity hormone leptin reduces eating by reducing meal size. In addition to white adipose tissue as a primary source, leptin is also released from the stomach. As peripheral leptin also crosses the blood brain barrier via a saturable transport mechanism, it may thus have access to multiple peripheral and central sites along the gut-brainstem neuraxis that are important in the control of meal size [30]. In support of this idea, LRb mRNA has been identified in rodent and human nodose ganglion neurons, and, coincidentally, many of these neurons co-express CCK receptor mRNA [31, 32]. Peripheral administration of leptin increased neurophysiological activity in gastric branch vagal afferents and in nodose ganglion neurons supplying the stomach and duodenum [33]. These neurons are also excited by local application of CCK, and are synergistically activated by combined administration of leptin and CCK. Furthermore, local administration of leptin along the upper gastrointestinal tract via the celiac artery dose-dependently reduced eating by reducing meal size, and potentiated the satiation effects of peripherally administered CCK [34]. The eating inhibition produced by celiac arterial leptin delivery was blocked by subdiaphragmatic vagotomy, and by systemic application of the neurotoxin capsaicin [35], which destroys a population of gut vagal afferents. Taken together, these data support a critical role for gut vagal afferents in the satiation effects of gastrointestinal leptin. Furthermore, they demonstrate that these afferents provide a shared channel, conveying multiple meal-related satiation signals to the brainstem, and suggesting that the synergistic interactions of peripheral satiation stimuli and leptin to limit meal size require an intact subdiaphragmatic vagus.

However, these data do not exclude the possibility that the caudal brainstem also contains neural substrates important for the synergistic eating inhibitory actions of leptin and post-oral satiation signals. LRb mRNA and immunoreactivity has been demonstrated in multiple rat and mouse brainstem nuclei important in the control of food intake and meal size, including the NTS and the DMX [36]. Signaling studies demonstrate that leptin induces increases in STAT3 phosphorylation, an index of functional leptin signaling, in DMX and NTS, supporting direct activation of these neuronal populations via LRb [37].

There is consistent and growing evidence that the gut recipient medial NTS (mNTS) is a critical central integrator of leptin and gut satiation signals. Central administration of leptin increased the eating inhibitory potency of gastric balloon distension, gastric nutrient loads, and peripheral CCK [37, 38]. Such combinations of leptin and meal-related postoral stimuli also synergistically increased the expression of c-Fos protein, a marker of neuronal activation, in medial and commisural NTS subnuclei [37, 38]. Central third ventricular administration of leptin at eating inhibitory doses potentiated the neurophysiological excitatory responses to gastric distending loads in mNTS neurons [39] (although see Williams et al. [40] for demonstrations of hyperpolarizing effects of leptin in gastric-recipient, possibly second-order, mNTS neurons). Gut-recipient mNTS neurons express FRb, and gastric distension paired with brainstem leptin administration increased not only mNTS c-Fos expression, but also stimulated mNTS STAT3 phosphorylation [37]. Furthermore, co-administration of subthreshold brainstem leptin doses paired with subthreshold gastric distending loads significantly reduced short-term food intake, while neither stimulus did so when presented alone [37]. Taken together, these data support the ability of mNTS neurons to integrate leptin signaling and peripheral meal-related gut vagal afferent signals important in the control of meal size.

Intrinsic Brainstem Signaling Pathways

A second advance in our understanding of caudal brainstem function in the control of food intake comes from the identification and characterization of neuropeptides and their receptors intrinsic to the brainstem. The brainstem distribution of proopiomelanocortin (POMC) and melanocortin – 4 receptors (MC4R) provide a good example. POMC is expressed by two anatomically distinct neuronal populations in the CNS – the hypothalamic arcuate nucleus and the NTS. Alpha MSH, a cleavage product of POMC, is released by POMC neurons and can act at central MC4 receptor expressing neurons, localized to both the NTS and the DMX within the caudal brainstem [41].

Chronic, adenoviral assisted expression of POMC within the NTS markedly increased alpha-MSH levels, and produced a sustained reduction in food intake [42]. Accordingly, 4V and parenchymal DVC injections of the MC3/4 R agonist MTII acutely reduced food intake, while administration of the MC3/4 R antagonist SHU9119 increased eating [43, 44]. Each of these eating effects was mediated by a specific change in meal size, without altering meal frequency [39, 44, 45].

Fourth ventricular administration of MTII at doses that reduced meal size also dose-dependently activated phosphorylation of extracellular signal related kinases 1/2 (ERK1/2) in a subpopulation of NTS POMC neurons, and pharmacological inhibition of ERK phosphorylation by U0126 attenuates MTII-induced reductions in eating [45]. As brainstem melanocortin signaling affects eating by altering meal size, these data suggest that ERK phosphorylation in NTS POMC neurons plays important intracellular role in the control of meal size.

It is not surprising, then, that mNTS POMC neurons also respond to visceral afferent input. Electrical stimulation of visceral afferents of the solitary tract directly excited mNTS POMC neurons, and these responses were blocked by local brainstem application of CCK-1 receptor antagonists [46]. This suggests that CCK, acting at CCK-1 receptors on presynaptic, central vagal afferent glutamatergic terminals drives excitatory postsynaptic NTS responses. Furthermore, CCK brainstem CCK injections potentiate the excitatory effects of solitary tract electrical stimulation on mNTS POMC neurons [46]. Taken together, these data support the existence of direct synaptic connections between CCK-sensitive gut vagal afferents and mNTS POMC neurons. CCK satiation requires MC4R, as MC4R null mice do not reduce their food intake in response to CCK [47]. Peripheral eating inhibitory doses of CCK stimulated c-Fos expression and rapidly increased ERK1/2 phosphorylation specifically within mNTS POMC neurons [48]. In addition, brainstem application of U0126 blocked the eating inhibitory effects of peripheral CCK [45, 49]. These data suggest that ERK1/2 phosphorylation in mNTS POMC neurons mediates the brainstem processing of gut vagal afferent signals important in the negative feedback control of meal size.

Recent data demonstrate that MC4R mRNA is also expressed by rat nodose ganglion neurons, and that activation of presynpatic MC4R with alpha MSH or MTII excites post-synaptic NTS neurons receiving vagal afferent input [50]. Vagal afferent rhizotomy prior to brainstem alpha MSH or MTII application reduced glutamatergic excitation of post-synaptic NTS neurons, suggesting an important vagal afferent MC4R mediated input to the NTS [50]. Subdiaphragmatic gut vagal afferents do not appear to be a necessary mediator of the eating inhibitory effects of central MC4R agonists, as subdiphragmatic vagotomy did not block the ability of brainstem MTII administration to reduce food intake [51]. However, these results do not address the possibility that peripherally administered MTII could inhibit eating through its actions at supradiaphramatic vagal afferent MC4R, either at the level of the nodose ganglion, and/or at central vagal afferent terminals in the brainstem.

At least some of the alpha-MSH innervation of the NTS arises from within the brainstem itself, raising an additional possibility; an intrinsic brainstem POMC-MC4R signaling pathway [45]. In support of this idea, alpha-MSH immunoreactive fibers have been identified in close approximation to NTS neurons stimulated by peripheral eating inhibitory doses of CCK and by gastric distension [44]. As MC4R are required for normal CCK satiation (see Fan et al. [47]), this arrangement suggests a local neurocircuitry whereby NTS POMC neurons integrate post-oral feedback signals from peripheral visceral afferents with melanocortinergic signals of brainstem origin [44]. In this context, it would be important to know whether MC4R are expressed on NTS POMC neurons themselves. Available evidence supports excitatory actions of alpha-MSH at MC4R through the stimulatory Gs protein; consequently, NTS alpha-MSH release would be predicted to act at putative NTS POMC neuronal MC4R to stimulate additional alpha-MSH in a feedforward fashion. This feedforward excitation would potentiate the impact of gut vagal afferent glutamatergic stimulation of NTS neuronal activity, and further contribute to the inhibitory control of meal size.

Peripheral leptin administration at eating inhibitory doses activates STAT3 phosphorylation in ~50% of NTS POMC neurons [52], suggesting that brainstem melanocortin signaling may also mediate the eating inhibitory effects of leptin. Indeed, brainstem administration of SHU9119 blocked the ability of 4V leptin to inhibit food intake, revealing a local leptin-melanocortin circuit in the control of ingestion [53]. A role for brainstem MC4R agonism in the specific reduction of meal size produced by leptin has not been identified, but would be anticipated. In addition, brainstem leptin injections might be expected to potentiate the ability of central MC4R agonists to reduce eating by limiting meal size.

Descending Forebrain Influences on Brainstem Control of Meal Size

A third advance in understanding the brainstem integrative control of eating comes from studies that examine how descending hypothalamic projections modify the NTS neuronal processing of gut signals that limit meal size. This section will discuss recent progress in revealing NTS neurons that receive convergent gut vagal afferent meal termination signals and descending hypothalamic projections. Koletsky $LR^{fa/fa}$ rats lacking leptin receptors display increased eating and meal size, and are relatively insensitive to the eating inhibitory actions of peripheral CCK [54]. Adenoviral-associated expression of functional leptin receptors specifically in the arcuate/mediobasal hypothalamus (MBH) of these animals reduces their hyperphagia and meal size, and restores CCK satiation [54, 55]. In neurologically intact rats, mNTS neurons receive descending axonal projections from leptin-sensitive oxytocinergic PVN neurons [56]. In addition, oxytocin immunoreactive axons are densely localized in close proximity to mNTS neurons activated by eating inhibitory doses of peripheral CCK [57]. These data suggest a descending pathway through which leptin acts via oxytocinergic PVN neurons to drive mNTS neurons involved in the negative feedback control of meal size. Accordingly, third ventricular (3V) subthreshold doses of oxytocin receptor antagonist [d(CH(2))(5), Tyr (Me)(2), Orn(8)]-vasotocin] (OVT) that have no effect on eating when administered alone partially blocked the effects of central leptin on food intake. Furthermore, 3V OVT blocked the ability of leptin to potentiate the eating inhibitory and mNTS c-Fos responses to peripheral CCK [56]. Fourth ventricular OVT administration more potently stimulated eating than 3V injections, and 4V subthreshold doses of OVT attenuated the satiation effects of peripheral CCK [56]. Taken together, these data support the idea that gut-recipient NTS neurons integrate gut vagal afferent meal-related satiation signals with descending oxytocin signals to determine the control of meal size. This also suggests that descending PVN oxytocinergic neurons provide a tonic inhibition of eating at the level of the caudal brainstem. Consistent with this suggestion, local 4V injection of the oxytocin receptor antagonist OVT alone rapidly and dose dependently increased meal size [58].

There is growing evidence that forebrain melanocortin signaling modulates the brainstem processing of satiation signals as well. Neurons in the gut-recipient mNTS receive projections from MC4R expressing PVN neurons, and specific parenchymal injections of SHU9119 directly into the rat PVN, at doses that failed to stimulate eating when administered in the 4V, blocked the ability of peripheral CCK to reduce meal size [59]. These data support the idea that hypothalamic PVN projections modulate NTS satiation signals via PVN MC4R. Taken together, the data suggest a forebrain-hindbrain circuit where leptin binds to ObRb on arcuate neurons that project to MC4R expressing PVN neurons. These in turn project to mNTS sites to modulate the neural processing of gut signals that determine meal size.

Summary and Future Directions

The last 5 years has witnessed an intensified and successful focus on the brainstem processing of signals important in the negative feedback control of food intake. This focus is strongly supported by advances in the identification of gut vagal afferent meal-related signals and brainstem extra- and intracellular signaling pathways, as well as by improved experimental designs that permit more discrete assessment of descending forebrain influences on the brainstem processing of postoral, meal-related signals that limit meal size. The interpretive strength of the findings summarized in this paper would not be possible without the coordinated application of experimental approaches targeting multiple levels of analysis, from molecular to neurophysiological, immunocytochemical, pharmacological, genetic and, ultimately, behavioral. These coordinated approaches provide an experimental template that can applied to reveal how newly identified signaling molecules and receptors can be fruitfully integrated into our understanding of the brainstem control of ingestion in particular, and into the neural basis of eating in general. Results from studies of the role of central hypothalamic nutrient sensing in the control of food intake provide a recent example. Eating rapidly increases cerebrospinal fluid levels of the essential amino acid leucine, and leucine metabolism within the MBH activates forebrain PVN oxytocinergic neurons, as well as mNTS neurons. Increased leucine availability in the MBH also reduces food intake by reducing meal size, suggesting that it, too, engages an oxytocenergic PVN-mNTS circuit. Accordingly, 4V application of subthreshold doses of OVT blocks the ability of MBH leucine to reduce meal size [58]. These results, in turn, open the door to investigations of how the brainstem integrates hypothalamic signals arising from central nutrient sensors with meal-related, postoral gut afferent signals to determine meal size.

Acknowledgements

Supported by NIH DK R01 047208, Skirball Institute for Nutrient Sensing and the New York Obesity Research Center, NIH DK 02361.

References

1 Blevins JE, Baskin DG: Hypothalamic-brainstem circuits controlling eating; in Langhans W, Geary N: Frontiers in Eating and Weight Regulation. Forum Nutr. Basel, Karger, 2010, vol 63, pp 133–140.

2 Glatzle J, et al: Expression of 5-HT3 receptors in the rat gastrointestinal tract. Gastroenterology, 2002; 123:217–226.

3 Raybould H: Primary afferent response to signals in the intestinal lumen. J Physiol 2001;530:343.

4 Savastano DM, Carelle M, Covasa M: Serotonin-type 3 receptors mediate intestinal Polycose- and glucose-induced suppression of intake. Am J Physiol Regul Integr Comp Physiol 2005;288:R1499–R1508.

5 Glatzle J, et al: Chylomicron components activate duodenal vagal afferents via a cholecystokinin A receptor-mediated pathway to inhibit gastric motor function in the rat. J Physiol 2003;550:657–664.

6 Hayes MR, et al: 5-HT3 receptors participate in CCK-induced suppression of food intake by delaying gastric emptying. Am J Physiol Regul Integr Comp Physiol 2004;287:R817–R823.

7 Hayes MR, Covasa M: Dorsal hindbrain 5-HT3 receptors participate in control of meal size and mediate CCK-induced satiation. Brain Res 2006; 1103:99–107.

8 Wu XY, et al: Neurochemical phenotype of vagal afferent neurons activated to express C-FOS in response to luminal stimulation in the rat. Neuroscience 2005;130:757–767.

9 Czaja K, Ritter RC, Burns GA:Vagal afferent neurons projecting to the stomach and small intestine exhibit multiple N-methyl-D-aspartate receptor subunit phenotypes. Brain Res 2006;1119:86–93.

10 Czaja K, Ritter RC, Burns GA: N-methyl-D-aspartate receptor subunit phenotypes of vagal afferent neurons in nodose ganglia of the rat. J Comp Neurol 2006;496:877–885.

11 Slattery JA, et al: Potentiation of mouse vagal afferent mechanosensitivity by ionotropic and metabotropic glutamate receptors. J Physiol 2006;577:295–306.

12 Treece BR, Ritter RC, Burns GA: Lesions of the dorsal vagal complex abolish increases in meal size induced by NMDA receptor blockade. Brain Res 2000;872:37–43.

13 Covasa M, Ritter RC, Burns GA: NMDA receptor blockade attenuates CCK-induced reduction of real feeding but not sham feeding. Am J Physiol Regul Integr Comp Physiol 2004;286:R826–R831.

14 Berthoud HR, et al: Food-related gastrointestinal signals activate caudal brainstem neurons expressing both NMDA and AMPA receptors. Brain Res 2001;915:143–154.

15 Hung CY, et al: Hindbrain administration of NMDA receptor antagonist AP-5 increases food intake in the rat. Am J Physiol Regul Integr Comp Physiol 2006;290:R642–R651.

16 Gillespie BR, Burns GA, Ritter RC: NMDA channels control meal size via central vagal afferent terminals. Am J Physiol Regul Integr Comp Physiol 2005;289:R1504–R1511.

17 Guard DB, et al: Blockade of hindbrain NMDA receptors containing NR2 subunits increases sucrose intake. Am J Physiol Regul Integr Comp Physiol 2009;296:R921–R928.

18 Guard DB, et al: NMDA NR2 receptors participate in CCK-induced reduction of food intake and hindbrain neuronal activation. Brain Res 2009;1266:37–44.

19 Wan S, Browning KN: D-glucose modulates synaptic transmission from the central terminals of vagal afferent fibers. Am J Physiol Gastrointest Liver Physiol 2008;294:G757–G763.

20 Wan S, Browning KN: Glucose increases synaptic transmission from vagal afferent central nerve terminals via modulation of 5-HT3 receptors. Am J Physiol Gastrointest Liver Physiol 2008;295:G1050–G1057.

21 Drucker DJ, Biologic actions and therapeutic potential of the proglucagon-derived peptides. Nat Clin Pract Endocrinol Metab 2005;1:22–31.

22 Bucinskaite V, et al: Receptor-mediated activation of gastric vagal afferents by glucagon-like peptide-1 in the rat. Neurogastroenterol Motil 2009, p 978-e78.

23 Nishizawa M, et al: The hepatic vagal reception of intraportal GLP-1 is via receptor different from the pancreatic GLP-1 receptor. J Auton Nerv Syst 2000;80:14–21.

24 Ruttimann EB, et al: Intrameal hepatic portal and intraperitoneal infusions of glucagon-like peptide-1 reduce spontaneous meal size in the rat via different mechanisms. Endocrinology 2009;150:1174–1181.

25 Goke R, et al: Distribution of GLP-1 binding sites in the rat brain: evidence that exendin-4 is a ligand of brain GLP-1 binding sites. Eur J Neurosci, 1995;7:2294–2300.

26 Vrang N, et al: Gastric distension induces c-Fos in medullary GLP-1/2-containing neurons. Am J Physiol Regul Integr Comp Physiol 2003;285:R470–R478.

27 Parkinson JR, et al: Differential patterns of neuronal activation in the brainstem and hypothalamus following peripheral injection of GLP-1, oxyntomodulin and lithium chloride in mice detected by manganese-enhanced magnetic resonance imaging (MEMRI). Neuroimage 2009;44:1022–1031.

28 Hayes MR, Skibicka KP, Grill HJ: Caudal brainstem processing is sufficient for behavioral, sympathetic, and parasympathetic responses driven by peripheral and hindbrain glucagon-like-peptide-1 receptor stimulation. Endocrinology 2008;149:4059–4068.

29 Hayes MR, Bradley L, Grill HJ: Endogenous hindbrain glucagon-like peptide-1 receptor activation contributes to the control of food intake by mediating gastric satiation signaling. Endocrinology 2009;150:2654–2659.

30 Banks WA: Blood-brain barrier as a regulatory interface; in Langhans W, Geary N: Frontiers in Eating and Weight Regulation. Forum Nutr. Basel, Karger, 2010, vol 63, pp 102–110.

31 Burdyga G, et al: Expression of the leptin receptor in rat and human nodose ganglion neurones. Neuroscience, 2002;109:339–347.

32 Buyse M, et al: Expression and regulation of leptin receptor proteins in afferent and efferent neurons of the vagus nerve. Eur J Neurosci 2001;14:64–72.

33 Peters JH, Ritter RC, Simasko SM: Leptin and CCK selectively activate vagal afferent neurons innervating the stomach and duodenum. Am J Physiol Regul Integr Comp Physiol 2006;290:R1544–R1549.

34 Peters JH, Simasko SM, Ritter RC: Modulation of vagal afferent excitation and reduction of food intake by leptin and cholecystokinin. Physiol Behav 2006;89:477–485.

35 Peters JH, et al: Leptin-induced satiation mediated by abdominal vagal afferents. Am J Physiol Regul Integr Comp Physiol 2005;288:R879–R884.

36 Scott MM, et al: Leptin targets in the mouse brain. J Comp Neurol 2009;514:518–532.

37 Huo L, et al: Leptin and the control of food intake: neurons in the nucleus of the solitary tract are activated by both gastric distension and leptin. Endocrinology, 2007;148:2189–2197.

38 Moran TH, Aja S, Ladenheim EE: Leptin modulation of peripheral controls of meal size. Physiol Behav 2006;89:511–516.
39 Azzara AV, Sokolnicki JP, Schwartz GJ: Central melanocortin receptor agonist reduces spontaneous and scheduled meal size but does not augment duodenal preload-induced feeding inhibition. Physiol Behav 2002;77:411–416.
40 Pardini AW, et al: Distribution of insulin receptor substrate-2 in brain areas involved in energy homeostasis. Brain Res 2006;1112:169–178.
41 Kishi T, et al: Expression of melanocortin 4 receptor mRNA in the central nervous system of the rat. J Comp Neurol 2003;457:213–235.
42 Li G, et al: Melanocortin activation of nucleus of the solitary tract avoids anorectic tachyphylaxis and induces prolonged weight loss. Am J Physiol Endocrinol Metab 2007;293:E252–E258.
43 Williams DL, et al: Behavioral processes underlying the intake suppressive effects of melanocortin 3/4 receptor activation in the rat. Psychopharmacology (Berl), 2002;161:47–53.
44 Zheng H, et al: Brain stem melanocortinergic modulation of meal size and identification of hypothalamic POMC projections. Am J Physiol Regul Integr Comp Physiol 2005; 289:R247–R258.
45 Sutton GM, et al: Melanocortinergic modulation of cholecystokinin-induced suppression of feeding through extracellular signal-regulated kinase signaling in rat solitary nucleus. Endocrinology 2005; 146:3739–3747.
46 Appleyard SM, et al: Proopiomelanocortin neurons in nucleus tractus solitarius are activated by visceral afferents: regulation by cholecystokinin and opioids. J Neurosci 2005;25:3578–3585.
47 Fan W, et al: Cholecystokinin-mediated suppression of feeding involves the brainstem melanocortin system. Nat Neurosci 2004;7:335–336.
48 Babic T, et al: Phenotype of neurons in the nucleus of the solitary tract that express CCK-induced activation of the ERK signaling pathway. Am J Physiol Regul Integr Comp Physiol 2009;296:R845–R854.
49 Sutton GM, Patterson LM, Berthoud HR: Extracellular signal-regulated kinase 1/2 signaling pathway in solitary nucleus mediates cholecystokinin-induced suppression of food intake in rats. J Neurosci 2004;24:10240–10247.
50 Wan S, et al: Presynaptic melanocortin-4 receptors on vagal afferent fibers modulate the excitability of rat nucleus tractus solitarius neurons. J Neurosci 2008;28:4957–4966.
51 Williams DL, Kaplan JM, Grill HJ: The role of the dorsal vagal complex and the vagus nerve in feeding effects of melanocortin-3/4 receptor stimulation. Endocrinology 2000;141:1332–1337.
52 Ellacott KL, Halatchev IG, Cone RD: Characterization of leptin-responsive neurons in the caudal brainstem. Endocrinology 2006;147:3190– 3195.
53 Skibicka KP, Grill HJ: Hindbrain leptin stimulation induces anorexia and hyperthermia mediated by hindbrain melanocortin receptors. Endocrinology 2009;150:1705–1711.
54 Morton GJ, et al: Leptin action in the forebrain regulates the hindbrain response to satiety signals. J Clin Invest 2005;115:703–710.
55 Morton GJ, et al: Arcuate nucleus-specific leptin receptor gene therapy attenuates the obesity phenotype of Koletsky (fa(k)/fa(k)) rats. Endocrinology 2003;144:2016–2024.
56 Blevins JE, Schwartz MW, Baskin DG: Evidence that paraventricular nucleus oxytocin neurons link hypothalamic leptin action to caudal brain stem nuclei controlling meal size. Am J Physiol Regul Integr Comp Physiol 2004; 287:R87–R96.
57 Blevins JE, et al: Oxytocin innervation of caudal brainstem nuclei activated by cholecystokinin. Brain Res 2003;993:30–41.
58 Blouet C, et al: Mediobasal hypothalamic leucine sensing regulates food intake through activation of a hypothalamus-brainstem circuit. J Neurosci 2009; 29:8302–8311.
59 Blevins JE, et al: Forebrain melanocortin signaling enhances the hindbrain satiety response to CCK-8. Am J Physiol Regul Integr Comp Physiol 2009; 296: R476–R484.

Gary J. Schwartz, PhD
Departments of Medicine and Neuroscience, Albert Einstein College of Medicine of Yeshiva University
1300 Morris Park Ave., Golding 501
Bronx, NY 10461 (USA)
Tel. +1 718 430 2263, Fax +1 718 430 2204, E-Mail gjschwar@aecom.yu.edu

Langhans W, Geary N (eds): Frontiers in Eating and Weight Regulation.
Forum Nutr. Basel, Karger, 2010, vol 63, pp 152–163

Gaining New Insights into Food Reward with Functional Neuroimaging

Marianne T. Neary · Rachel L. Batterham

Centre for Diabetes and Endocrinology, Department of Medicine, University College London, London, UK

Abstract

The notion that eating is intimately related to feelings of pleasure is not new. Indeed, in an environment characterised by many varied and palatable foods, hedonistic drives are likely to play a greater role in modulating food intake than homeostatic ones. Until recently however, a neurobiological account of the rewarding properties of food was lacking. The ability to reveal functional brain activity has been made possible with the advent of functional neuroimaging techniques such as electroencephalography (EEG), magnetoencephalography (MEG), positron emission tomography (PET) and most recently, functional magnetic resonance imaging (fMRI). Neuroimaging studies in fed and fasted, lean and obese, normal and pathological states have revealed variations in food-related reward processing. Eating is a multi-sensory experience and understanding the precise mechanisms by which food modulates reward circuits will be important in understanding the aetiology of obesity and eating disorders. Here we review the development of functional neuroimaging as a research tool and recent neuroimaging studies relating to food reward. In particular, we evaluate the ability of leptin and the gut hormones peptide YY3–36 and ghrelin to modulate activity in reward-related brain regions. Finally, we discuss the potential to use such information to guide development of pharmaceuticals, functional foods and life-style modifications.

> 'Nothing would be more tiresome than eating and drinking if God had not made them a pleasure as well as a necessity.'
>
> Voltaire (1694–1778)

Food and sex are essential for the survival of any species. Yet these behaviours are not simply a 'necessity' or rather, unconscious homeostatic drives; there is no doubt that hunger encourages us to eat, but when sated, few of us subsequently let the desert menu pass us by.

Stemming from Darwin's theory of evolution, the thrifty gene hypothesis [1] proposes that in an environment characterised by competition for food and mates, traits which drive obtainment of these have a selective advantage and are more likely to be passed on to the next generation. An adapted sense of hedonism that can be activated by food and sex is an obvious means of increasing our drive to engage in these behaviours. In contrast to adaptations such as a faster running speed or greater agility, which aid in obtaining homeostatic levels of food intake, hedonic desires encourage over-eating. Over-indulgence promotes the accumulation of excess fat reserves, thereby increasing survival through periods of famine. So successful has this adaptation been that, surrounded by many varied and palatable foods, we cannot resist the temptation to eat, even in the absence of impending famine. The genetic vestiges of evolutionary success have never been as evident as they are in our increasingly obese world.

Although Darwin has answered the question of 'why' we have hedonistic control of eating, the neurobiological basis of 'how' this is achieved is only just starting to be uncovered. In a society with plentiful food, hedonistic drives to eat are playing a more important role in over-consumption than homeostatic drives, despite research in the field of obesity largely focusing on the latter. Therefore, understanding 'how' is of prime importance for the development of effective strategies and therapeutics aimed at preventing and reducing obesity and other states of disordered eating.

Much insight into the control of eating, mainly homeostatic, has been gained from mechanistic studies in rodents [2]. Deciphering food-related reward processes on the other hand is complex, not least because eating represents a multi-sensory experience involving several brain regions. Functional neuroimaging provides unique insights into food-reward regulation by allowing the simultaneous assessment of brain activity with external stimuli, subjective appetite ratings, physiological and metabolic changes, and eating itself. Here we review the development of functional neuroimaging as a research tool, evaluate the ability of leptin and the gut hormones peptide YY3–36 (PYY3–36) and ghrelin to modulate activity in reward-related brain regions, and discuss the potential to use such information to guide development of pharmaceuticals, functional foods and life-style modifications. Further aspects of the progress in this fast-moving field are described in two other chapters in this volume [3, 4]

A Brief Timeline of Functional Neuroimaging

For millennia, man has aspired to understand the workings of the mind. In the 5th century BC, Socrates utilised the notion of 'introspection': looking into one's mind through one's experiences. Over 2,000 years later, one of the founding figures of modern psychology, Wilhelm Wundt, attempted to use introspection as a means of experimentally dissecting the human mind into basic elements [5]. Highly trained assistants were given an experience such as a ticking metronome to reflect upon and self-report their feelings and motivations. Introspection was a dominant method in development of experimental psychology, but in the early 20th century it was increasingly criticised for being too subjective, unable to reveal unconscious activity preceding a given action or feeling, and unable to localise the origins of such processes.

It was not until the 1920s that brain activity could be measured objectively. Hans Berger noticed that electrodes on his daughters head detected a stronger electrical signal the more difficult arithmetic she was doing. He concluded that the frequency of the wave pattern correlated to brain activity. These experiments provided the basis of the modern-day electroencephalogram (EEG) [6]. Stemming from this, in the 1960s David Cohen developed and measured the first magnetoencephalogram (MEG) [7]. A MEG is founded on the same basic principles as the EEG, the difference being that it detects the magnetic component of neuronal electrical activity (see [8] for a comprehensive account of EEG and MEG).

The concept of positron emission tomography (PET) was conceived in the 1950s by David Kuhl and Roy Edwards [9]. However, this was not employed on human subjects until the late 1970s [10]. In PET, a positron-emitting radioisotope is intravenously injected. The most commonly used isotope is a non-metabolizable analogue of glucose, fludeoxyglucose (^{18}F-FDG) which accumulates in brain regions with high metabolic activity. Similarly, ^{15}O-containing water can be used, which increases in brain regions of high blood flow. It is assumed that metabolic activity and blood flow are good correlates for neuronal activity. When the radioisotope decays it emits a positron, which travels a short distance before colliding with an electron. Annihilation of the two particles produces two gamma rays, emitted in opposite directions. These are detected by a scintillator surrounding the individual. By comparing the timing of the gamma ray arrival and position on the scintillator, the signal origin can be localised to a specific brain region. Additionally, by using radioisotopes that are ligands of neurotransmitter receptors or transporters, PET can measure the differential activation of specific neural pathways (for an in-depth review of PET, see [11]).

Another breakthrough for functional neuroimaging came in 1990, when Seiji Ogawa [12] discovered that deoxyhaemoglobin could be distinguished from oxyhaemoglobin in magnetic resonance imaging (MRI). This enabled the visualisation of differential blood flow through brain regions at a given moment in time on the backdrop of a highly detailed MRI image. MRI is a non-invasive technique employing strong magnetic fields and radio wave pulses. In the presence of the magnetic field, mobile hydrogen nuclei in biological molecules and water align with the direction of the field. Radio-wave pulses matching the resonance frequency of hydrogen nuclei excite them into a higher energy state. When the nuclei return to their original state, they release energy that distorts the magnetic field. This distortion forms the basis of the MRI signal. These working principles mean that the MRI signal is affected by the magnetic properties of hydrogen nuclei, conferred by their parent molecules and molecular complexes. Deoxy- and oxyhaemoglobin can be distinguished because their hydrogen nuclei are paramagnetic and diamagnetic, respectively. The greater the neuronal activity of a brain region, the greater its blood flow and relative concentration of oxyhaemoglobin which is detected as a stronger MRI signal (for a review, see [13]). In 1991, Kenneth Kwong [14] used this technique, referred to as functional

MRI (fMRI), to assess brain function in humans. There are now many more innovations of fMRI in addition to 'BOLD' (blood-oxygen level dependent), including diffusion-based fMRI, which detects changes in cell shape induced by neuronal activity [15] and arterial spin labelling, which non-invasively alters the magnetic properties of proximal blood flow to the brain [16].

Table 1 summarises the common functional neuroimaging techniques and their advantages and disadvantages. For simplicity, the rest of this chapter will focus on studies using fMRI.

fMRI and Food Reward

Food-related cues evoke activity changes in regions of the brain implicated chiefly in homeostatic regulation, as well as those associated with reward, such as the orbital frontal cortex (OFC), insula and dorsal striatum [17] (fig. 1). Activity changes in these regions are often more pronounced in the fasted compared to fed state, concomitant with elevated ratings of hunger and the pleasantness of food [18–20] These findings provide a neurobiological basis for the more rewarding properties of food as well as increased motivation to seek more palatable foods in the fasted state. (For more detailed discussions of the brain regions involved in reward processing, the reader is referred to other chapters in this volume [3, 4, 21].)

Differential activation is also dependent on body weight: a greater number of reward-associated brain regions show food-related activity changes in obese compared to lean individuals [22–24]. This discrepancy could lead to the aberrant motivation of some obese individuals to consume more and highly palatable foods. Whether this is a cause or consequence of obesity remains to be elucidated as the relative infancy of functional neuroimaging as a research tool has meant that prospective studies examining potential causality have not yet been undertaken.

Variation has also been demonstrated in individuals with disordered eating. Abnormal activation of the anterior cingulate cortex was demonstrated in patients with bulimia nervosa [25]. In Prader-Willi syndrome, where individuals invariably display hyperphagia and obesity, orbitofrontal cortex and hypothalamic dysfunction were observed [26, 27]. Awry activity changes were also seen in the caudate region of recovering anorexia nervosa sufferers [28].

What underlies the modulation of these reward circuits? An obvious assumption is that we are driven to eat some foods due to orosensory signals such as taste. It is likely that these signals are important in reward-based eating [29]. However, the finding that mice lacking taste receptors still developed a strong preference for sucrose solutions [30] suggests the involvement of additional factors.

In the following sections, we focus on how eating-related hormones influence reward processing. These hormones vary in fed and fasted states, lean and obese individuals and also in pathological disorders of eating [31] and could in part underlie

Table 1. Summary of the major attributes and drawbacks of major functional neuroimaging techniques

	Basis of function	Advantages	Disadvantages
EEG	Electrodes attached to the scalp directly detect electric fields generated by neuronal activity	Electrical activity is direct correlate of neuronal activity High temporal resolution Portable and cheap equipment Non-invasive	Detection limited to cortical activity
MEG	Similar to EEG, except MEG detects the magnetic component of electrical neuronal activity	As above In addition, MEG signal is less distorted than EEG by resistive properties of skull	As above
PET	A radioisotope is injected into the peripheral circulation of the subject By detecting its radioactive decay, the differential concentration in brain regions can be visualised Positron-emitting glucose or water is used to show differences in metabolic rate or blood flow respectively, both markers of neuronal activation	High spatial resolution Radioisotopes can be used which bind to specific neurotransmitter receptors or transporters	Low temporal resolution Indirectly measures neuronal activity Costly Invasive injection of exogenous agent Radiation dose to subject Size and weight constraints sometimes prohibit scanning of obese individuals
fMRI	The BOLD technique uses the different magnetic properties of deoxy- and oxyhaemoglobin to detect changes in blood flow, a marker of neuronal activation	Higher spatial resolution than PET Non-invasive	Low temporal resolution Indirectly measures neuronal activity Costly Size and weight constraints sometime prohibit scanning in obese individuals Brainstem structures difficult to visualise without cardiac and respiratory gating

BOLD = Blood-oxygen level dependent; EEG = electroencephalogram; fMRI = functional magnetic resonance imaging; MEG = magnetoencephalogram; PET = positron emission tomography.

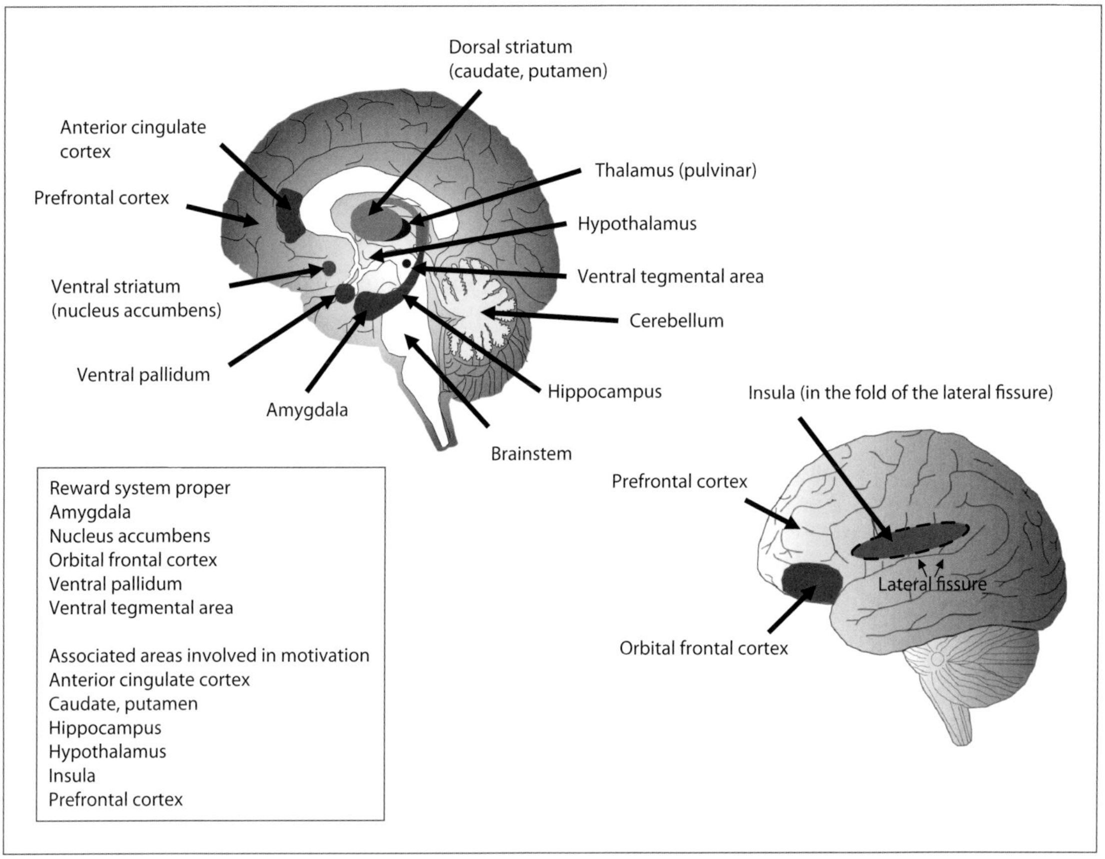

Fig. 1. Schematic of some of the brain regions known to play a role in reward processing.

the variations described above. So far, fMRI studies have evaluated the brain-specific effects of three eating related hormones: leptin, PYY3–36 and ghrelin, demonstrating that they do indeed play a role in the hedonistic control of eating.

Eating-Related Hormones and Food Reward

Leptin

Three separate fMRI studies have examined the relationship between the plasma levels of the adipocycte hormone, leptin, and brain activation: two were conducted in individuals with congenital leptin deficiency [32, 33] and one in obese subjects [34]. In the latter study by Rosenbaum and colleagues, obese subjects (BMI >30) underwent caloric restriction until their baseline weight was reduced by 10%. They were then given just enough food to maintain this weight. Reduction and maintenance of the new body weight was associated with a decrease in circulating leptin concentrations. After stabilising at their new weights, the subjects received placebo or twice daily injections of leptin that brought plasma leptin back to its baseline level. This was done in a cross-over design; subjects received both treatments in sequence so that they each acted as their own controls. Neural responses to food images were assessed with fMRI before weight loss and after weight loss, with and without leptin replacement. The authors found that after weight loss, activity changed in several areas implicated in reward processing, including the amygdala, cingulate cortex, hippocampus and frontal cortex. Many of these changes in activity were reversed by leptin replacement. These findings offer a possible explanation for the difficulties experienced in maintaining weight loss; decreased leptin levels result in enhanced activation of reward circuits, therefore driving behaviour towards eating, especially foods which are highly palatable. Moreover, this study suggests a potential therapeutic role for leptin-replacement therapy in maintaining weight loss after dieting.

PYY3–36

The anorectic gut hormone PYY3–36 is released by L cells of the distal gut in proportion to calorie content and macronutrient quality of a meal [35]. In 2007, Batterham et al. [36] undertook a double-blind placebo crossover study in which they infused fasted healthy volunteers with either placebo (saline) or PYY3–36 to mimic the physiological PYY3–36 concentrations observed in the fed state. In this way, contributions of taste and other meal-related sensory experiences were absent. They demonstrated that PYY3–36 modulated activity in both homeostatic, hypothalamus and brainstem, and non-homeostatic regions including the amygdala, OFC, ventral tegmental area (VTA), ventral striatum and insular cortex. They also found that in the presence of low circulating PYY3–36 (saline infusion), activity in the hypothalamus predicted the subjects subsequent ad libitum caloric intake, whereas in the presence of high, post-prandial PYY3–36 levels activity in the OFC predicted their subsequent caloric intake (fig. 2). Furthermore, there was a negative correlation between OFC activation and meal pleasantness ratings. Taken in concert, these findings suggest that PYY3–36 enhances discrimination of rewarding foods and diminishes reward-motivated drives to eat in the fed state. Therefore the reduced fasting and post-prandial PYY levels in obesity [37] and elevated fasting levels in anorexia nervosa [38] may contribute to differential desire for food and therefore food intake in these individuals.

Ghrelin

Ghrelin, released from X/A cells in the gastric mucosa during fasting, is the only gut hormone known to stimulate appetite and meal initiation [39]. In 2008, Malik et al.

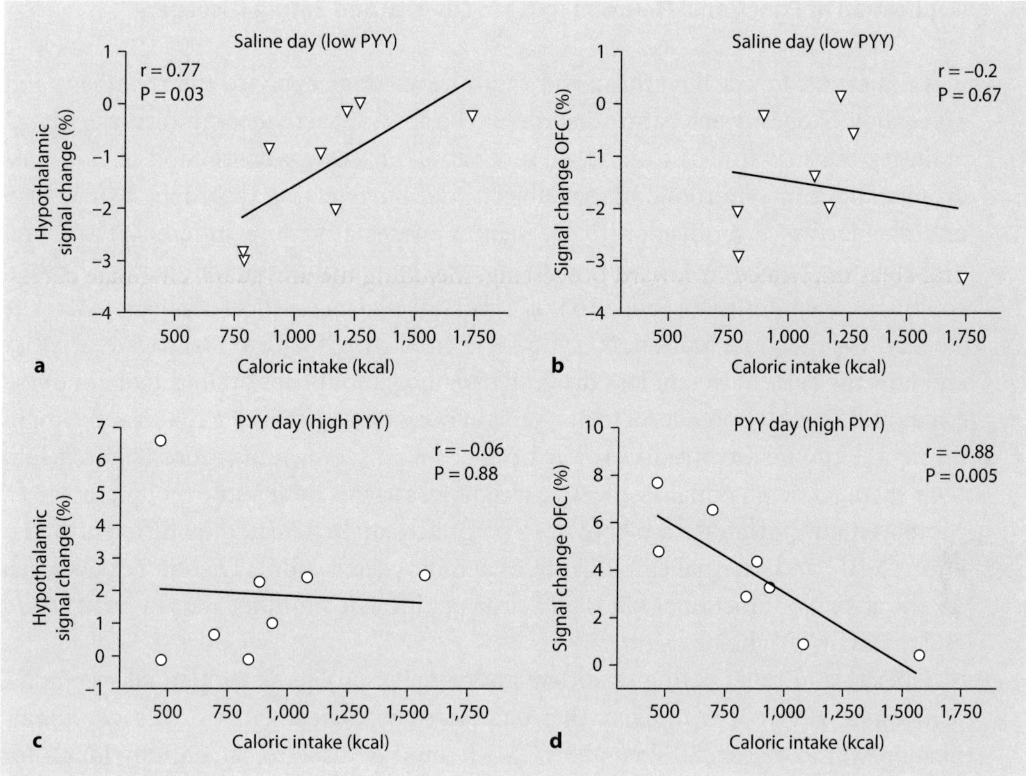

Fig. 2. Post-prandial levels of PYY switch the regulation of food intake from homeostatic hypothalamic control to hedonic. Data from a fMRI study conducted by Batterham et al. [36] showing the correlation of hypothalamic and OFC activity with subsequent food intake after saline and PYY3–36 infusion.

[40] infused ghrelin or placebo to 20 non-obese male volunteers 3 h after consuming a standard meal. They demonstrated that ghrelin administration altered neural responses to images of food in regions such as the amydala, VTA, OFC, anterior insula and dorsal striatum. In addition, self-reports of hunger were increased in the ghrelin-infused state and varied positively with activity in the amygdala, OFC and pulvinar, which is consistent with increased desire for palatable foods in the fasted state. Unlike the PYY3–36 study, ghrelin did not induce activity changes in the ventral striatum, suggesting that these hormones have overlapping but distinct effects on reward processing. This study offers a neurological basis for the voracious hyperphagia, particularly of energy-dense foods, observed in individuals with Prader-Willi syndrome who display fasting and post-prandial hyperghrelinism [41].

Application of Functional Neuroimaging in Obesity and Eating Disorders

It is important to ask how these and future revelations apply to the treatment and prevention of obesity and eating disorders. The aberrant activity of reward-associated brain regions may underlie increased motivation for energy-dense food seen in syndromic and non-syndromic obese subjects and binge-eating disorders. It may also underlie decreased motivation to eat seen in anorexia nervosa sufferers. Therefore, external manipulation of these reward circuits, by pharmacological means or other, could alter their desire for food. To this effect, the pharmaceutical company, Merck & Co. Inc. (Whitehouse Station, N.J., USA), is employing fMRI to investigate whether and how the current weight loss drug, sibutramine, modulates brain activity in obese males [42]. The clear effects of leptin and gut hormones such as PYY3–36 and ghrelin on neural circuits involved in reward processing also warrants more research into their therapeutic potential as effective weight loss agents or appetite modulators [31]. Moreover, gut hormones are known to be preferentially released by different nutrients [43–45] and external stimuli such as aerobic exercise [46]. Therein lies potential for the advent of functional foods and clear-cut lifestyle modifications in order to aid and maintain weight loss.

Obesity and other eating disorders are complex problems with equally complex aetiologies. Hence it is unlikely that one pharmacological agent or lifestyle modification will work for all. Potentially, fMRI could be used to screen individuals for aberrant reward processing. Application of a patient-tailored dose of a specific drug or intervention in susceptible individuals could alleviate their awry motivation for food.

Future of Functional Neuroimaging in Appetite Regulation Research

The power of neuroimaging to investigate the neural control of eating will no doubt increase with improved methodology; enhancing image contrast in fMRI by peripheral administration of exogenous agents such as manganese in rodent studies for example [47]. Simultaneous use of the different neuroimaging techniques will combine their advantageous properties; the high spatial resolution of fMRI combined with PET could more accurately map specific neurochemical pathways underlying food reward. Addition of EEG and MEG, would enhance temporal resolution of these changes [48].

The advent of functional imaging technology has already contributed to our knowledge of the neural control of eating in humans, but many questions remain unanswered. The inter-play between hedonistic and homeostatic controls of food intake is far from clear, for example. How other non-homeostatic factors such as emotion and cognition influence food intake and their interaction with reward processing should also be evaluated in future studies. Reward has recently been described as having 3 main components: learning, wanting and liking [49], each of these components arising from distinguishable underlying neurological substrates [50]. Studies evaluating the reward-modulating effects of stimuli such as the hormones considered here should also consider how they differentially affect each of these processes; liking a particular food in a presented image is different from wanting that food item. Moreover, many variables await assessment for their effects on reward-circuitry: other metabolic hormones, lack of sleep, acute and chronic stress, etc.

Importantly, despite advances in elucidating the effects of various physiological and pathological states on neural activity, we still have little knowledge of how these brain regions are functionally organised. Psychological constructs such as appetite, reward and cravings are still poorly or variably operationalized and often used inappropriately as explanations. It would be futile to simply produce vast quantities of data on neural activation without knowing or appreciating the functional implications of these regions. Great progress is likely to be made by combining mechanistic studies in rodents with functional imaging and the multi-disciplinary expertise of psychologists, endocrinologists and neuroscientists alike. And with commitment and the application of functional neuroimaging and other powerful techniques, ongoing research endeavours will help bring us one step closer to understanding and solving many eating-related disorders.

References

1 Neel JV: Diabetes mellitus: a 'thrifty' genotype rendered detrimental by 'progress'? Am J Hum Genet 1962;14:353–362.

2 Speakman J, Hambly C, Mitchell S, Krol E: The contribution of animal models to the study of obesity. Lab Anim 2008;42:413–432.

3 Kringelbach ML, Stein A: Cortical mechanisms of human eating; in Langhans W, Geary N (eds): Frontiers in Eating and Weight Regulation. Forum Nutr. Basel, Karger, 2010, vol 63, pp 164–175.

4 Stice E, Dagher A: Genetic variation in dopaminergic reward in humans; in Langhans W, Geary N (eds): Frontiers in Eating and Weight Regulation. Forum Nutr. Basel, Karger, 2010, vol 63, pp 176–185.

5 Wundt W: Grundzüger der physiologischen Psychologie. Leipzig, Engelman, 1874.

6 Berger H: Über das Elektroenkephalogramm des Menschen. Arch Psychiatr Nervenkrankh 1929;87:527–570.

7 Cohen D: Magnetoencephalography: evidence of magnetic fields produced by alpha-rhythm currents. Science 1968;161:784–786.

8 Niedermeyer E, Da Silva FHL: Electroencephalography: Basic Principles, Clinical Applications and Related Fields, ed 5. Philadelphia, Lippincott Williams & Wilkins, 2004.

9 Kuhl DE, Edwards RQ: Image separation radioisotope scanning. Radiology 1963;80:653–661.

10 Reivich M, Kuhl D, Wolf A, Greenberg J, Phelps M, Ido T, Casella V, Fowler J, Hoffman E, Alavi A, Som P, Sokoloff L: The [18F]fluorodeoxyglucose method for the measurement of local cerebral glucose utilization in man. Circ Res 1979;44:127–137.

11 Mittra E, Quon A: Positron emission tomography/computed tomography: the current technology and applications. Radiol Clin N Am 2009;47:147–160.

12 Ogawa S, Lee TM, Nayak AS, Glynn P: Oxygenation-sensitive contrast in magnetic resonance image of rodent brain at high magnetic fields. Magn Reson Med 1990;14:68–78.

13 Gore JC: Principles and practice of functional MRI of the human brain. J Clin Invest 2003;112:4–9.

14 Kwong KK, Belliveau JW, Chesler DA, Goldberg IE, Weisskoff RM, Poncelet BP, Kennedy DN, Hoppel BE, Cohen MS, Turner R: Dynamic magnetic resonance imaging of human brain activity during primary sensory stimulation. Proc Natl Acad Sci USA 1992;89:5675–5679.

15 Le Bihan D, Urayama S, Aso T, Hanakawa T, Fukuyama H: Direct and fast detection of neuronal activation in the human brain with diffusion MRI. Proc Natl Acad Sci USA 2006;103:8263–8268.

16 Petersen ET, Zimine I, Ho YCL, Golay X: Noninvasive measurement of perfusion: a critical review of arterial spin labelling techniques. Br J Radiol 2006;79:688–701.
17 Killgore WD, Young AD, Femia LA, Bogorodzki P, Rogowska J, Yurgelun-Todd DA: Cortical and limbic activation during viewing of high- versus low-calorie food. Neuroimage. 2003;19;1381–1394.
18 Fuhrer D, Zysset S, Stumvoll M: Brain activity in hunger and satiety: an exploratory visually stimulated fMRI study. Obesity (Silver Spring) 2008;16: 945–950.
19 Siep N, Roefs A, Roebroeck A, Havermans R, Bonte ML, Jansen A: Hunger is the best spice: an fMRI study of the effects of attention, hunger and calorie content on food reward processing in the amygdala and orbitofrontal cortex. Behav Brain Res 2009;198: 149–158.
20 Smeets PA, de Graaf C, Stafleu A, can Osch MJ, Nievelstein RA van der Grond J: Effect of satiety on brain activation during chocolate tasting in men and women. Am J Clin Nutr 2006;83;1297–1305.
21 Langhans W, Geary N: Overview of the physiological control of eating; in Langhans W, Geary N (eds): Frontiers in Eating and Weight Regulation. Forum Nutr. Basel, Karger, 2010, vol 63, pp 9–53.
22 Killgore WD, Yurgelun-Todd DA: Body mass predicts orbitofrontal cortex activity during visual representations of high calorie foods. Neuroreport 2005;16:859–863.
23 Rothemund Y, Preuschhof C, Bohner G, Bauknecht HC, Klingebiel R, Flor H, Klapp BF: Differential activation of the dorsal striatum by high-calorie visual food stimuli in obese individuals. Neuroimage 2007;37:410–421.
24 Stoeckel LE, Weller RE, Cook EW, Twieg DB, Knowlton RC, Cox JE: Widespread reward-system activation in obese women in response to pictures of high-calorie foods. Neuroimage 2008;41:636–647.
25 Frank GK, Wagner A, Achenbach S, McConaha C, Skovira K, Aizenstein H, Carter CS, Kaye WH: Altered brain activity in women recovered from bulimic-type eating disorders after a glucose challenge: a pilot study. Int J Eat Disord 2006;39:76–79.
26 Dimitropoulos A, Schultz RT: Food-related neural circuitry in Prader-Willi syndrome: response to high- versus low-calorie foods. J Autism Dev Disord 2008;38:1642–1653.
27 Miller JL, James GA, Goldstone AP, Couch JA, He G, Driscoll DJ, Liu Y: Enhanced activation of reward mediating prefrontal regions in response to food stimuli in Prader-Willi syndrome. J Neurol Neurosurg Psychiatry 2007;78:615–619.
28 Wagner A, Aizenstein H, Venkatraman VK, Fudge J, May JC, Mazurkewicz L, Frank GK, Bailer UF, Fischer L, Nguyen V, Carter C, Putnam K, Kaye WH: Altered reward processing in women recovered from anorexia nervosa. Am J Psychiatry 2007; 164:1842–1849.
29 Rolls ET: Taste, olfactory, and food texture processing in the brain, and the control of food intake. Physiol Behav 2005;85:45–56.
30 de Araujo IE, Oliveira-Maia AJ, Sotnikova TD, Gainetdinov RR, Caron MG, Nicolelis MA, Simon SA: Food reward in the absence of taste receptor signaling. Neuron 2008;57:930–941.
31 Neary MT, Batterham RL: Gut hormones: implications for the treatment of obesity. Pharmacol Ther 2009;124:44–56.
32 Farooqi IS, Bullmore E, Keogh J, Gillard J, O'Rahilly S, Fletcher PC: Leptin regulates striatal regions and human eating behavior. Science 2007;317:1355.
33 Baicy K, London ED, Monterosso J, Wong ML, Delibasi T, Sharma A, Licinio J: Leptin replacement alters brain response to food cues in genetically leptin-deficient adults. Proc Natl Acad Sci USA 2007;104:18276–18279.
34 Rosenbaum M, Sy M, Pavlovich K, Leibel RL, Hirsch J: Leptin reverses weight loss-induced changes in regional neural activity responses to visual food stimuli. J Clin Invest 2008;118:2583–2591.
35 Neary MT, Batterham RL: Peptide YY: Food for thought. Physiol Behav 2009;97:616–619.
36 Batterham RL, ffytche DH, Rosenthal JM, Zelaya FO, Barker GJ, Withers DJ, Williams SC: PYY modulation of cortical and hypothalamic brain areas predicts feeding behaviour in humans. Nature 2007;450:106–109.
37 le Roux CW, Batterham RL, Aylwin SJB, Patterson M, Borg CM, Wynne KJ, Kent A, Vincent RP, Gardiner J, Ghatei MA, Bloom SR: Attenuated peptide YY release in obese subjects is associated with reduced satiety. Endocrinology 2006;147:3–8.
38 Misra M, Miller KK, Tsai P, Gallagher K, Lin A, Lee N, Herzog DB, Klibanski A: Elevated peptide YY levels in adolescent girls with anorexia nervosa. J Clin Endocrinol Metab 2006;91:1027–1033.
39 Tschöp M, Smiley DL, Heiman ML: Ghrelin induces adiposity in rodents. Nature 2000;407:908–913.
40 Malik S, McGlone F, Bedrossian D, Dagher A: Ghrelin modulates brain activity in areas that control appetitive behavior. Cell Metab 2008;7:400–409.
41 Feigerlova E, Diene G, Conte-Auriol F, Molinas C, Gennero I, Salles JP, Arnaud C, Tauber M: Hyperghrelinemia precedes obesity in Prader-Willi syndrome. J Clin Endocrinol Metab 2008;93:2800–2805.

42 A Functional Magnetic Resonance Imaging (fMRI) Study in Overweight and Obese Men. Merck & Co. Inc., 2009. (Accessed June 22nd, 2009, at http://clinicaltrials.gov/ct2/show/NCT00914212.)

43 Batterham RL, Heffron H, Kapoor S, Chivers JE, Chandarana K, Herzog H, Le Roux CW, Thomas EL, Bell JD, Withers DJ: Critical role for peptide YY in protein-mediated satiation and body-weight regulation. Cell Metabol 2006;4:223–233.

44 Douglas BR, Jansen JB, Tham RT, Lamers CB: Coffee stimulation of cholecystokinin release and gallbladder contraction in humans. Am J Clin Nutr 1990;52:553–556.

45 Helou N, Obeid O, Azar ST, Hwalla N: Variation of postprandial PYY 3–36 response following ingestion of differing macronutrient meals in obese females. Ann Nutr Metab 2008;52:188–195.

46 Broom DR, Batterham RL, King JA, Stensel DJ: The influence of resistance and aerobic exercise on hunger, circulating levels of acylated ghrelin and peptide YY in healthy males. Am J Physiol Regul Integr Comp Physiol 2009;296;29–35.

47 Parkinson JR, Chaudhri OB, Bell JD: Imaging appetite-regulating pathways in the central nervous system using manganese-enhanced magnetic resonance imaging. Neuroendocrinology 2009;89:121–30.

48 Ritter P, Villringer A: Simultaneous EEG-fMRI. Neurosci Biobehav [R] 2006;30:823–838.

49 Berridge KC, Robinson TE: Parsing reward. Trends Neurosci 2003;26:507–513.

50 Berridge KC, Robinson TE, Aldridge JW: Dissecting components of reward: 'liking', 'wanting', and learning. Curr Opin Pharmacol 2009;9:65–73.

Dr. Rachel L. Batterham
Centre for Diabetes and Endocrinology, Department of Medicine, University College London
Rayne Building, 5 University Street
London WC1E 6JJ (UK)
Tel. +44 2076790991, Fax +44 2076796583, E-Mail r.batterham@ucl.ac.uk

Langhans W, Geary N (eds): Frontiers in Eating and Weight Regulation.
Forum Nutr. Basel, Karger, 2010, vol 63, pp 164–175

Cortical Mechanisms of Human Eating

Morten L. Kringelbach[a,b] · Alan Stein[a]

[a]Department of Psychiatry, University of Oxford, Oxford, UK; [b]Aarhus University, Aarhus, Denmark

Abstract

The hedonic component of eating is an underexplored topic within neuroscience, which is surprising given its importance for our survival and general well-being, as well as the obvious links to obesity and eating disorders. Based on findings from neuroimaging, this review gives an overview of the established principles, neural mechanisms and functional neuroanatomy of the primate and human brain processing systems involved in controlling eating. Four main processing principles underlying these processes are discussed: (1) motivation-independent discriminative processing of identity and intensity, (2) formation of learning-dependent multimodal sensory representations, (3) reward representations using mechanisms including selective satiation, and (4) representations of hedonic experience, monitoring/learning or direct behavioural change.

Maintaining nutrient homeostasis by eating is closely linked to the evolutionary imperatives of survival and procreation, and it has been argued that the rewards and pleasures associated with eating must play key roles in the brain [1–3]. Consummatory behaviour is rewarding in itself and is, along with basic homeostatic regulation, hardwired in even brainless species. However, the challenges of controlling eating are much greater for mammals, which must maintain a stable body temperature in a wide variety of hostile climates. This in turn requires intricate neural circuits which, at least in humans, include mechanisms associated with a sensation of subjective pleasure.

The relative sophistication of foraging in higher primates compared to other mammals indicates that significant parts of their large brains are dedicated to the required motivational, emotional and cognitive processing. It has been proposed that some of these cortical networks evolved for the more advanced aspects of eating-related behaviour have been recycled and have come to underlie other higher cognitive functions [4].

Human eating relies on complex brain processing that, among other functions, obtains stable sensory information, evaluate desirability and select the appropriate behaviour (fig. 1). Part of this processing is linked to basic homeostatic regulation [5], which has been elucidated in great detail in animal models with mammals including humans sharing many sub-cortical circuits and molecules, as outlined in this book [6–10] and elsewhere [11, 12].

Human eating is not, however, controlled only by homeostatic processes, as illustrated by our easy overindulgence on sweet foods beyond homeostatic needs and the epidemic proportions of obesity, which has become a major health problem [13]. Instead, the control of human eating relies on the interaction between homeostatic regulation and pleasure and reward, with important modulation by other factors including genetics [14], circadian rhythms [15], reproductive status [16] and social factors. This complex sub-cortical and cortical processing involves higher-order processing, including learning, memory, planning and prediction, and gives rise to conscious experience of not only the sensory properties of the food (such as the identity, intensity, temperature, fat content and viscosity) but also the affective valence elicited by the food (including, most importantly, the pleasure experienced) [6, 17].

This chapter reviews the evidence linking cortical regions of the human brain to aspects of human eating and the associated subjective experiences. This evidence was obtained through neurophysiological findings in other primates and rodents and was then further elaborated in human neuroimaging and neuropsychology studies. Other aspects of contemporary neuroimaging work are described in two other chapters in this volume [18, 19].

Motivation and the Taste System

Reward and hedonic processing are closely linked to motivation and emotion. A useful distinction has been proposed between two aspects of reward: hedonic impact and incentive salience, where the former refers to the 'liking' or pleasure related to the reward, e.g. the experiences of eating, and the latter to the 'wanting' or desire to obtain the reward [20, 21]. Clearly, brain regions implicated in hedonic assessment must receive salient information about stimulus identity from the primary and secondary sensory cortices. Thus, neuroimaging offers a powerful way to investigate both the 'liking' and 'wanting' components in the human brain. For example, one way to investigate 'liking' is to correlate hedonic ratings taken throughout a human neuroimaging experiment with changes in brain activity [22–24]. This allows for a unique window on the hedonic processes evaluating the pleasantness of salient stimuli.

Eating-related behaviour involves crucial decisions, where the brain must compare and evaluate the predicted reward value of various behaviours. This processing can be complex, as the estimations will vary in quality depending on the sampling rate of the behaviour and the variance of reward distributions. It is hard to provide a reliable estimate of the reward value of a food that appears to be highly desirable and is high in nutritional value but is only rarely available and varies significantly in

Fig. 1. Processes and substrates for controlling eating. Eating is a complex process. From foraging and procurement ('behaviour') to expectation, selection, and decision making ('brain'), it relies heavily on many lower and higher brain processes. These are in turn modulated by many factors ('modulation'). In this review we concentrate on cortical stages of brain processing related to the 'behaviour' and 'brain' processes. Note that there are also many important digestive, gut-brain, autonomic and metabolic interactions which are starting to be elucidated.

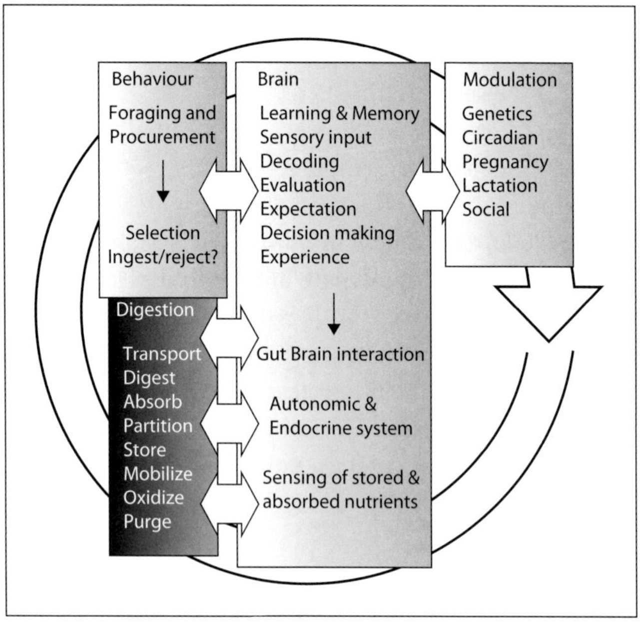

quality. This raises a classic problem in animal learning, how to optimize behaviour such that the amount of exploration is balanced with the amount of exploitation, where exploration is the time spent sampling the outcome of different behaviours and exploitation is the time spent using existing behaviours with known reward values. Eating-related behaviours have to be precisely controlled for another reason: erroneous evaluation of the sensory properties of foods, if it leads to the decision to swallow toxins, microorganisms or non-food objects, can be fatal. Humans and other animals have therefore developed elaborate eating-related behaviours to balance conservative risk-minimising and life-preserving strategies (exploitation) with occasional novelty seeking (exploration) in the hope of discovering new, valuable sources of nutrients [25].

Pleasure and hedonic processing in general are central to this balancing act between exploitation and exploration. The evidence from neuroimaging studies has linked regions of the human brain – and in particular the orbitofrontal cortex – to various aspects of eating and especially to the representation of the subjective pleasantness of foods [4].

From Sensory Processing to Hedonic Experience of Food

All the five classic senses (vision, hearing, smell, taste, and touch) are involved in the control of eating. In addition, other sensory receptors contribute, such as those in the digestive tract which are sensitive to gastric distension, or those in the circulatory system that are sensitive to changes in blood pressure or carbon dioxide gas in the blood. Nevertheless, perhaps the two most important senses involved in eating are smell and taste. In the following it is shown how they interact to facilitate and modulate decision making and hedonic experience. Four processing principles have been proposed for the interaction between sensory and hedonic processing in humans: (1) motivation-independent discriminative processing of identity and intensity; (2) formation of learning-dependent multimodal sensory representations; (3) reward representations using state-dependent mechanisms including selective satiation, and (4) representations of hedonic experience, monitoring/learning or direct behavioural change [26].

Motivation-Independent Processing of Identity and Intensity

Neuroimaging and human brain lesions studies indicate that the primary taste area in humans is located in the anterior insula/frontal operculum [27–30]. This is also consistent with the findings in primates [31, 32]. The largest fMRI study of taste processing to date used forty datasets from thirty-eight right-handed subjects with (1) identical delivery of the taste stimuli, (2) the same control procedure in which a tasteless solution was delivered after every taste stimulus, and (3) event-related interleaved designs [31]. A total of eight unimodal and six multimodal taste stimuli (oral stimuli that produce typically taste, olfactory and somatosensory stimulation) ranging from pleasant to unpleasant were used in the four experiments. Stringent analysis of taste activity across the forty datasets revealed three cortical activity foci in response to the main effects of taste in the human brain (fig. 2). Bilateral activity in the anterior insular/frontal opercular cortex was found with a slightly stronger response on the right side. This slight asymmetry in bilateral taste processing fits with an early meta-analysis of gustatory responses gathered from neuroimaging studies suggesting that the preponderance of activity peaks to taste fall in the right hemisphere [29]. Taste stimuli also produced activity in the medial caudal orbitofrontal cortex, which is likely to coincide with the secondary taste cortex.

As with taste stimuli, neuroimaging studies of pure olfactory stimuli reveal dissociable brain areas for motivation-independent representations of reinforcer identity and for hedonic representations: representations of olfactory identity occur in primary olfactory cortices [33–38], distinct from hedonic representations in other brain areas including the orbitofrontal cortex.

In general, the experiments in humans and non-human primates clearly demonstrate that the primary sensory areas for taste and smell are not modulated by motivational state, and that hedonic processing is generally thought to occur in higher-order, multi-modal areas such as the orbitofrontal cortex.

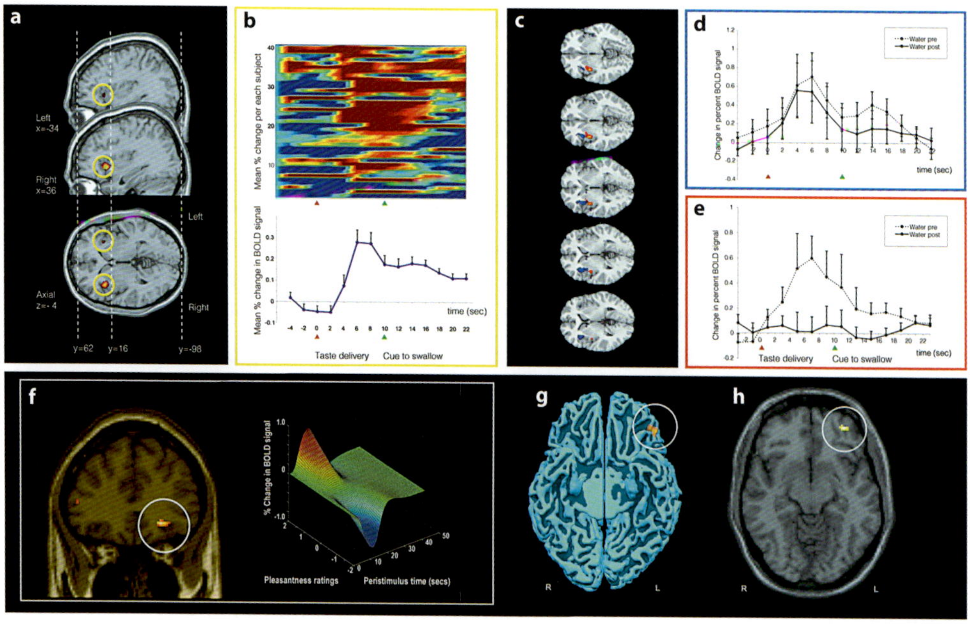

Fig. 2. Principles of taste processing and pleasure. Dissociable parts of the human primary taste cortex in the insula/operculum are involved in different aspects of food intake. **a** Neuroimaging has located the primary human taste cortex in bilateral anterior insular/frontal opercular cortices (yellow circles) with peak MNI coordinates of [x, y, z: 38, 20, –4] and [x, y, z: –32, 22, 0] (top two sagittal and axial slices) [31]. This is based on data from 40 right-handed subjects over four experiments with eight unimodal and six multimodal taste stimuli ranging from pleasant to unpleasant and found, in concordance with data from non-human primates. **b** The time course of blood oxygen-level dependent (BOLD) activity in right primary taste cortex is shown for all forty subjects (top), and averaged across all (bottom). The ordinate shows each of the individual datasets, while the abscissa shows the time from the taste delivery (time zero, orange triangle) through when the subject is cued to swallow after 10 s (green triangle). The magnitude of the activation is encoded in a scale of percent change from red (activation) to blue (deactivation) relative to the mean. The data were averaged for each dataset for all the different taste minus the tasteless solution. The average time course across the forty datasets with the standard error of the mean is shown at the bottom. **c** Dissociable functions of two different parts of the insula and their time courses of activity. Axial slices showing the extent of primary taste cortex (in blue) which is *not* modulated by thirst. In contrast, the region of right mid-insula (in red) is modulated by thirst [23]. **d** Time courses of activation extracted from a cluster in right primary taste cortex with respect to the delivery of tasteless (throughout the experiment) water with the response shown separately for the pre-satiety and post-satiety states. **e** Similar time courses from a cluster in mid-insula cortex showing significantly modulatory effects of water between states when satiated and thirsty. **f** From this initial decoding, the signal are further processed in the human orbitofrontal cortex. Links between subjective pleasure and activation of a mid-anterior site in orbitofronal cortex have been demonstrated using subjective pleasantness ratings of foods in a selective-satiety study [24], **g** supra-additive effects of combining the umami tastants monosodium glutamate and inosine monophosphate [22], and **h** supra-additive effects of combining strawberry odour with sucrose taste solution [23].

Formation of Learning-Dependent Multimodal Sensory Representations

In addition integrating information from taste and smell receptors, decisions about eating also integrate somatosensory information from the oral and nasal cavities. This sensory information includes temperature, viscosity, fat contents, pungency and irritation and is mediated by a large variety of neural systems. This information is integrated by a hierarchical system distributed form the brainstem to the cerebral cortex and is made available for the crucial higher-order decision of ingestion or rejection of a potentially poisonous food [39]. Neuroimaging investigations of such multimodal integration have found that one of the central brain regions is the human orbitofrontal cortex. Auditory [40], gustatory [29], olfactory [41], somatosensory [42] and visual [43] inputs, as well as information from the visceral sensory system [44], all produce activity in the human orbitofrontal cortex. This is consistent with neurophysiological recordings finding that the non-human primate orbitofrontal cortex receives input from all of the five senses [45]. These sensory inputs enter the orbitofrontal cortex mostly through its posterior parts. The interaction between taste and smell revealed by with neuroimaging is found in slightly more anterior parts of the orbitofrontal cortex and nearby agranular insula [24, 28, 46].

A good example of multimodal integration is how subjective olfactory experience appears different depending on whether a smell reaches the nasal cavity through the nose (orthonasal) or mouth via the posterior nares of the nasopharynx (retronasal) [47]. These are likely to be related to differences in somatosensory influences (e.g. mastication). Several neuroimaging studies have found corresponding differences in cortical activity patterns between ortho- and retronasal olfaction in the orbitofrontal cortex and related brain regions [22, 48, 49].

Reward Representations of Sensory Stimuli

In contrast to the motivation-independent representations of uni- and multimodal reinforcer identities, neuroimaging studies have found that affective valence is encoded in a network of brain regions. For example, in a neuroimaging taste study, a dissociation was found between the brain regions responding to the intensity of the taste and its affective valence [50]. Brain regions responding to intensity regardless of valence included the cerebellum, pons, middle insular cortex, and amygdala, whereas valence-specific responses were observed in the orbitofrontal cortex, with the right caudolateral orbitofrontal cortex responding preferentially to pleasant compared to unpleasant taste, irrespective of intensity. Another neuroimaging study found that the subjective ratings of taste pleasantness (but not intensity) correlated with activity in the medial orbitofrontal cortex and in the anterior cingulate cortex [23]. Moreover, in the same study it was found that activity in the medial orbitofrontal cortex and a region of mid-insula correlated with subjective pleasantness ratings of water during thirst and subsequent replenishment.

Further evidence of neural correlates of subjective experience of pure taste was found in an experiment investigating true taste synergism, which is the phenomenon whereby the intensity of a taste is dramatically enhanced by adding minute doses of another taste. The results of this neuroimaging experiment showed that the strong subjective enhancement of the pleasantness of umami taste that occurs when 0.005 M inosine 5′-monophosphate is added to 0.5 M monosodium glutamate (compared to both delivered separately) correlated with increased activity in an mid-anterior part of the orbitofrontal cortex (fig. 2g) [22].

Several neuroimaging olfaction studies have found dissociable encoding of the intensity and pleasantness of olfactory stimuli, with the intensity encoded in the amygdala and nearby regions, and the pleasantness correlated with activity in the medial orbitofrontal cortex and anterior cingulate cortex [35–37]. This is consistent with studies that have found that hedonic judgments are correlated with activity in the medial orbitofrontal cortex [33] and that the unpleasantness of aversive odours correlates with activity in the lateral orbitofrontal cortex [38]. Furthermore, it has been found that the orbitofrontal cortex represents the sensory-specific decrease of smell [34], which is clear evidence that the reward value of olfactory stimuli is represented in the orbitofrontal cortex.

Other strong evidence for the role of the orbitofrontal cortex in the representation of the reward value of olfactory stimuli comes from an appetitive conditioning neuroimaging experiment in which brain activity related to two arbitrary visual stimuli was measured both before and after olfactory devaluation, i.e. where the reward value of a stimulus is temporarily devalued by e.g. eating it to satiety [51]. In the amygdala and the orbitofrontal cortex, responses evoked by a predictive target stimulus decreased after devaluation, whereas responses to the non-devalued stimulus were maintained. It would thus appear that differential activity in the amygdala and the orbitofrontal cortex encodes the current value of reward representations accessible to predictive cues.

This evidence is compatible with studies in non-human primates with lesions of the orbitofrontal cortex. In one study, lesioned monkeys responded normally to associations between food and conditioners but failed to modify their behaviour to the cues when the incentive value of the food was reduced [52], and, in another study, where lesioned monkeys displayed altered food preferences [53]. Similarly, monkey with unilateral lesions of the orbitofrontal cortex on one side and of the basolateral part of the amygdala on the other side displayed disrupted stimulus devaluation in a procedure in which the incentive value of a food was reduced by satiation on that specific food [54].

Representations of Hedonic Experience

The evidence from neuroimaging studies of pure taste and smell cited above shows that the orbitofrontal cortex is consistently correlated with the subjective pleasantness ratings of the stimuli. Therefore, it is to be expected that studies using multi-modal combinations of taste and smell as well as state-dependent changes in pleasantness should find correlations between subjective pleasantness and activity in these brain regions. Compelling evidence that this is indeed the case comes from a selective-satiety neuroimaging study in which a region of the left orbitofrontal cortex showed not only a selective decrease in the reward value to the food eaten to satiety (and not to the food not eaten), but also a correlation with pleasantness ratings (see fig. 2f) [24]. This result indicates that the reward value of the taste, olfactory, and somatosensory components of a food are represented in the orbitofrontal cortex and, therefore, that the subjective pleasantness of food might be represented here.

Further evidence for the convergence of taste and smell hedonic processing comes from a study investigating the non-specific satiation effects of chocolate (with both olfactory and gustatory components) which found a correlation between the decrease in pleasantness and activity in the orbitofrontal cortex [55]. Another multimodal study investigating the link between olfaction and vision found activity in the anterior orbitofrontal cortex for semantically congruent trials [56]. Finally, when investigating the synergistic enhancement of a matched taste and retronasal smell, it was again found that a region of the orbitofrontal cortex was significantly active (fig. 2) [46]. This region was located very near to the region of the orbitofrontal cortex active by the synergistic combinations of umami described above [22].

It is an open but interesting question whether the orbitofrontal cortex and perhaps even sub-regions thereof are both necessary and sufficient for the experience of sensory and social pleasure. The evidence from psychosurgery studies of last century is not illuminating because of the usually crude psychological measurements and because the lack of neuroimaging or careful post-mortem investigations meant that the surgical lesions were not adequately described. One study of patients with relatively circumscribed lesions suggests that white-matter lesions that disconnect the orbitofrontal cortex can lead to serious emotional changes [57]. Direct tests of anhedonia linked to lesions of the orbitofrontal cortex have, however, not been carried out.

Conclusions

Nutritional homeostasis is essential to sustain life, and the brain's natural reward systems contributes fundamentally to meeting this need [1]. The evidence reviewed here suggests that the control of eating-related behaviours, eating itself, and the subjective experiences associated with it rely on cortical processing in humans and other primates. We propose four main processing principles: (1) motivation-independent discriminative processing of identity and intensity; (2) formation of learning-dependent multimodal sensory representations; (3) reward representations using mechanisms including selective satiation, and (4) representations of hedonic experience,

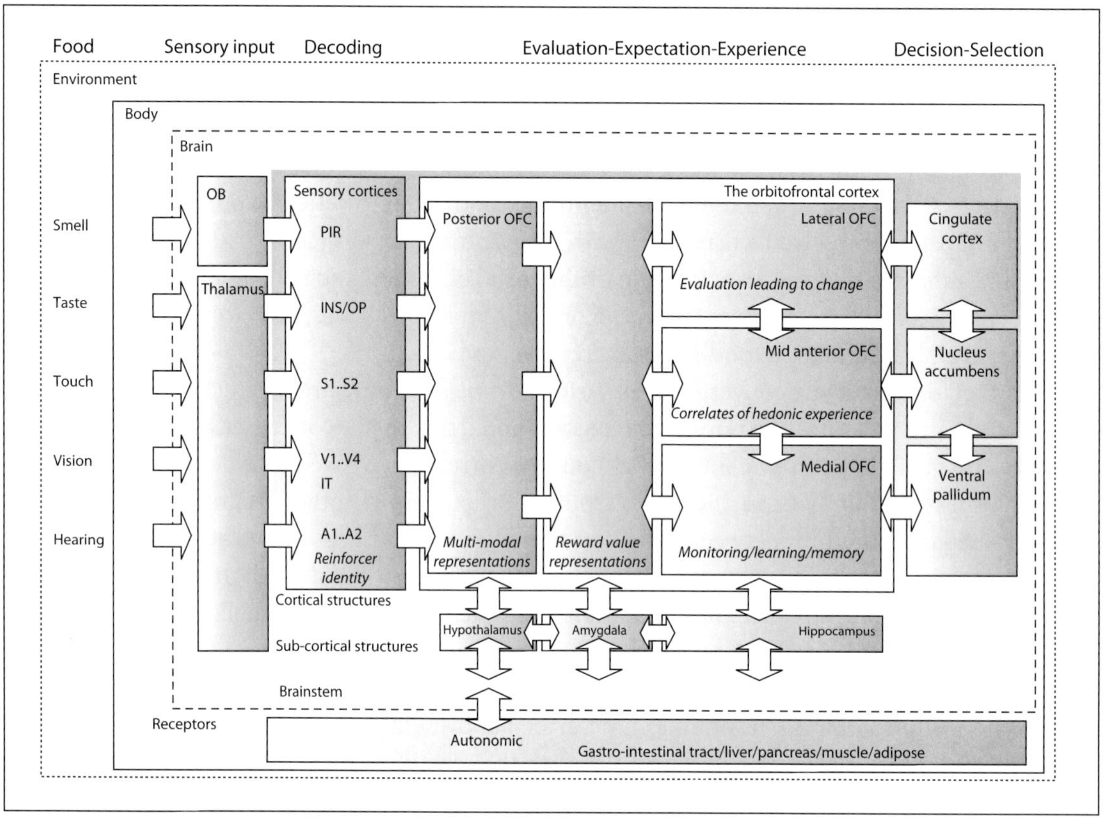

Fig. 3. Converging brain pathways in the brain processing of food stimuli. The figure summarises the interactions with environment to procure suitable food sources with a special focus on the role of the orbitofrontal cortex. Potential food sources are identified on the basis of the sensory input, which through the appropriate receptors are relayed to the orbitofrontal cortex (only one hemisphere shown), where processing is taking place of evaluation, expectation, experience as well as decision and selection. Here the input is processed in the primary sensory cortices via the thalamus (except for olfaction) and made available for pattern-association between primary (e.g. taste) and secondary (e.g. visual) reinforcers. Stimulus sensory identities are then processed for multimodal perceptual integration in the posterior orbitofrontal cortex. Hedonic reward value is represented in more anterior parts of orbitofrontal cortex, from where it can then be used to influence subsequent behaviour (in lateral parts of the anterior orbitofrontal cortex with connections to anterior cingulate cortex), stored for valence learning/memory (in medial parts of the anterior orbitofrontal cortex) and made available for subjective hedonic experience (in mid-anterior orbitofrontal cortex). There is multiple modulatory brain-loops with other important structures such as the nucleus accumbens, ventral pallidum, hippocampus, amygdala and hypothalamus, as well as modulation with autonomic input from the gut [5]. V1, V2, V4 = Primary and secondary visual areas; SS = somatosensory cortex (3,1,2); A1..A2 = auditory cortex; INS/OP = insular cortex/frontal operculum; IT = inferior temporal visual cortex; PIR = piriform cortex; OB = olfactory bulb; OFC = orbitofrontal cortex.

monitoring/learning or direct behavioural change. All but the first of these processes rely to a large extent on the orbitofrontal cortex. In line with earlier proposals [4] a possible model is proposed which implements these computational principles for the interaction between sensory and hedonic systems in the human brain (fig. 3). This simplified model of the orbitofrontal cortex begins to address the basic principles of eating is controlled in the human brain. More research into these cortical mechanisms is likely not only to further elucidate the workings of this network but, perhaps, also to illuminate larger questions, such as the neural correlates of subjective experience.

Acknowledgements

This research is supported by the TrygFonden Charitable Foundation and Wellcome Trust.

References

1 Kringelbach ML, Berridge KC: Pleasures of the Brain. New York, Oxford University Press, 2009.
2 Berridge KC, Kringelbach ML: Affective neuroscience of pleasure: reward in humans and animals. Psychopharmacology 2008;199:457–480.
3 Kringelbach ML: The Pleasure Center: Trust Your Animal Instincts. New York, Oxford University Press, 2009.
4 Kringelbach ML: Food for thought: hedonic experience beyond homeostasis in the human brain. Neuroscience 2004;126:807–819.
5 Batterham RL, Ffytche DH, Rosenthal JM, et al: PYY modulation of cortical and hypothalamic brain areas predicts feeding behaviour in humans. Nature 2007;450:106–109.
6 Langhans W, Geary N: Overview of the physiological control of eating; in Langhans W, Greary N (eds): Frontiers in Eating and Weight Regulation. Forum Nutr. Basel, Karger, 2010, vol 63, pp 9–53.
7 Blevins JE, Baskin DG: Hypothalamic brainstem circuits controlling eating and energy homeostasis; in Langhans W, Greary N (eds): Frontiers in Eating and Weight Regulation. Forum Nutr. Basel, Karger, 2010, vol 63, pp 133–140.
8 Bouret SG: Development of hypothalamic neural networks controlling appetite; in Langhans W, Geary N (eds): Frontiers in Eating and Weight Regulation. Fourm Nutr. Basel, Karger 2010, vol 63, pp 84–93.
9 Moran TH: Hypothalamic nutrient sensing; in Langhans W, Greary N (eds): Frontiers in Eating and Weight Regulation. Forum Nutr. Basel, Karger, 2010, vol 63, pp 94–101.
10 Schwartz GJ: Brainstem integrative function in the central nervous system of food intake; in Langhans W, Greary N (eds): Frontiers in Eating and Weight Regulation. Forum Nutr. Basel, Karger, 2010, vol 63, pp 141–151.
11 Berthoud HR, Morrison C: The brain, appetite, and obesity. Annu Rev Psychol 2008;59:55–92.
12 Saper CB, Chou TC, Elmquist JK: The need to feed: homeostatic and hedonic control of eating. Neuron 2002;36:199–211.
13 Kohn M, Booth M: The worldwide epidemic of obesity in adolescents. Adolesc Med 2003;14:1–9.
14 O'Rahilly S, Farooqi IS: Genetics of obesity. Phil Trans R Soc Lond 2006;361:1095–1105.
15 Ramsey KM, Marcheva B, Kohsaka A, Bass J: The clockwork of metabolism. Ann Rev Nutr 2007;27: 219–240.
16 Eckel LA: Estradiol: a rhythmic, inhibitory, indirect control of meal size. Physiol Behav 2004;82:35–41.
17 Kringelbach ML. The human orbitofrontal cortex: linking reward to hedonic experience. Nat Rev Neurosci 2005;6:691–702.
18 Neary MT, Batterham RL: Gaining new insights into food reward with functional neuroimaging; in Langhans W, Greary N (eds): Frontiers in Eating and Weight Regulation. Forum Nutr. Basel, Karger, 2010, vol 63, pp 152–163.
19 Stice E, Dagher A: Genetic variation in dopamineric reward in humans; in Langhans W, Greary N (eds): Frontiers in Eating and Weight Regulation. Forum Nutr. Basel, Karger, 2010, vol 63, pp 176–185.
20 Berridge KC: Food reward: brain substrates of wanting and liking. Neurosci Biobehavl Rev 1996;20:1–25.

21 Berridge KC, Robinson TE: What is the role of dopamine in reward: hedonic impact, reward learning, or incentive salience? Brain Res Rev 1998;28:309–369.
22 De Araujo IET, Kringelbach ML, Rolls ET, Hobden P: The representation of umami taste in the human brain. J Neurophysiol 2003:313–319.
23 De Araujo IET, Kringelbach ML, Rolls ET, McGlone F: Human cortical responses to water in the mouth, and the effects of thirst. J Neurophysiol 2003;90: 1865–1876.
24 Kringelbach ML, O'Doherty J, Rolls ET, Andrews C: Activation of the human orbitofrontal cortex to a liquid food stimulus is correlated with its subjective pleasantness. Cerebral Cortex 2003;13:1064–1071.
25 Rozin P: Food Preference; in Smelser NJ, Baltes PB (eds): International Encyclopedia of the Social and Behavioral Sciences. Amsterdam, Elsevier, 2001.
26 Kringelbach ML: Cortical systems involved in appetite and food consumption; in Cooper SJ, Kirkham TC (eds): Appetite and Body Weight: Integrative Systems and the Development of Anti-Obesity Drugs. London, Elsevier, 2006, pp 5–26.
27 Kinomura S, Kawashima R, Yamada K, et al: Functional anatomy of taste perception in the human brain studied with positron emission tomography. Brain Res 1994;659:263–266.
28 Small DM, Jones-Gotman M, Zatorre RJ, Petrides M, Evans AC: Flavor processing: more than the sum of its parts. Neuroreport 1997;8:3913–3917.
29 Small DM, Zald DH, Jones-Gotman M, et al: Human cortical gustatory areas: a review of functional neuroimaging data. Neuroreport 1999;10:7–14.
30 O'Doherty J, Rolls ET, Francis S, Bowtell R, McGlone F: Representation of pleasant and aversive taste in the human brain. J Neurophysiol 2001;85:1315–1321.
31 Kringelbach ML, de Araujo IE, Rolls ET: Taste-related activity in the human dorsolateral prefrontal cortex. Neuroimage 2004;21:781–788.
32 Pritchard TC, Edwards EM, Smith CA, et al: Gustatory neural responses in the medial orbitofrontal cortex of the old world monkey. J Neurosci 2005;25:6047–6056.
33 Royet JP, Hudry J, Zald DH, et al: Functional neuroanatomy of different olfactory judgments. Neuroimage 2001;13:506–519.
34 O'Doherty J, Rolls ET, Francis S, et al: Sensory-specific satiety-related olfactory activation of the human orbitofrontal cortex. Neuroreport 2000;11: 893–897.
35 Rolls ET, Kringelbach ML, de Araujo IET: Different representations of pleasant and unpleasant odors in the human brain. Eur J Neurosci 2003;18:695–703.
36 Anderson AK, Christoff K, Stappen I, et al: Dissociated neural representations of intensity and valence in human olfaction. Nat Neurosci 2003;6: 196–202.
37 Gottfried JA, Deichmann R, Winston JS, Dolan RJ: Functional heterogeneity in human olfactory cortex: an event-related functional magnetic resonance imaging study. J Neurosci 2002;22:10819–10828.
38 Zald DH, Pardo JV: Emotion, olfaction, and the human amygdala: amygdala activation during aversive olfactory stimulation. Proc Natl Acad Sci USA 1997;94:4119–4124.
39 Grill HJ, Norgren R: The taste reactivity test. II. Mimetic responses to gustatory stimuli in chronic thalamic and chronic decerebrate rats. Brain Res 1978;143:281–297.
40 Frey S, Kostopoulos P, Petrides M: Orbitofrontal involvement in the processing of unpleasant auditory information. Eur J Neurosci 2000;12:3709–3712.
41 Zatorre RJ, Jones-Gotman M, Evans AC, Meyer E: Functional localization and lateralization of human olfactory cortex. Nature 1992;360:339–340.
42 Rolls ET, O'Doherty J, Kringelbach ML, Francis S, Bowtell R, McGlone F: Representations of pleasant and painful touch in the human orbitofrontal and cingulate cortices. Cerebral Cortex 2003;13:308–317.
43 Aharon I, Etcoff N, Ariely D, Chabris CF, O'Connor E, Breiter HC: Beautiful faces have variable reward value: fMRI and behavioral evidence. Neuron 2001; 32:537–551.
44 Critchley HD, Mathias CJ, Dolan RJ: Fear conditioning in humans: the influence of awareness and autonomic arousal on functional neuroanatomy. Neuron 2002;33:653–663.
45 Rolls ET: The Brain and Emotion. Oxford, Oxford University Press, 1999.
46 De Araujo IET, Rolls ET, Kringelbach ML, McGlone F, Phillips N: Taste-olfactory convergence, and the representation of the pleasantness of flavour, in the human brain. Eur J Neurosci 2003;18:2059–2068.
47 Pierce J, Halpern BP: Orthonasal and retronasal odorant identification based upon vapor phase input from common substances. Chem Senses 1996; 21:529–543.
48 Cerf-Ducastel B, Murphy C: fMRI activation in response to odorants orally delivered in aqueous solutions. Chem Senses 2001;26:625–637.
49 Small DM, Gerber JC, Mak YE, Hummel T: Differential neural responses evoked by orthonasal versus retronasal odorant perception in humans. Neuron 2005;47:593–605.

50 Small DM, Gregory MD, Mak YE, Gitelman D, Mesulam MM, Parrish T: Dissociation of neural representation of intensity and affective valuation in human gustation. Neuron 2003;39:701–711.
51 Gottfried JA, O'Doherty J, Dolan RJ: Encoding predictive reward value in human amygdala and orbitofrontal cortex. Science 2003;22:1104–1107.
52 Butter CM, Mishkin M, Rosvold HE: Conditioning and extinction of a food-rewarded response after selective ablations of frontal cortex in rhesus monkeys. Exp Neurol 1963;7:65–75.
53 Baylis LL, Gaffan D: Amygdalectomy and ventromedial prefrontal ablation produce similar deficits in food choice and in simple object discrimination learning for an unseen reward. Exp Brain Res 1991; 86:617–622.
54 Baxter MG, Parker A, Lindner CC, Izquierdo AD, Murray EA: Control of response selection by reinforcer value requires interaction of amygdala and orbital prefrontal cortex. J Neurosci 2000; 20:4311–4319.
55 Small DM, Zatorre RJ, Dagher A, Evans AC, Jones-Gotman M: Changes in brain activity related to eating chocolate: from pleasure to aversion. Brain 2001; 124:1720–1733.
56 Gottfried JA, Dolan RJ: The nose smells what the eye sees: crossmodal visual facilitation of human olfactory perception. Neuron 2003;39:375–386.
57 Hornak J, Bramham J, Rolls ET, et al: Changes in emotion after circumscribed surgical lesions of the orbitofrontal and cingulate cortices. Brain 2003;126: 1671–1712.

Morten L. Kringelbach
Department of Psychiatry, University of Oxford, Warneford Hospital
Oxford OX3 7JX (UK)
Tel. +44 1865223784, Fax +44 1865226384, E-Mail Morten.Kringelbach@psych.ox.ac.uk

Langhans W, Geary N (eds): Frontiers in Eating and Weight Regulation.
Forum Nutr. Basel, Karger, 2010, vol 63, pp 176–185

Genetic Variation in Dopaminergic Reward in Humans

Eric Stice · Alain Dagher

Oregon Research Institute, Eugene, Oreg., USA

Abstract

Dopamine-based reward circuitry appears to play a role in encoding reward from eating and incentive sensitization, whereby cues associated with food reward acquire motivational value. Data suggest that low levels of dopamine D2 receptors and attenuated responsivity of dopamine-target regions (e.g. the striatum) to food and food cues are associated with elevated weight. There is mixed evidence that genotypes that appear to be associated with reduced signaling of dopamine circuitry, including DRD2, DRD4 and DAT, are correlated with obesity. In addition, there is emerging fMRI evidence that reduced responsivity in brain regions implicated in food reward increase risk for future weight gain among individuals who appear to be at genetic risk for attenuated dopamine signaling by virtue of DRD2 and DRD4 genotypes. However, it is vital for these relations to be replicated in larger, independent prospective studies and to use positron emission tomography to better characterize parameters of dopamine signaling, including dopamine receptor density, basal dopamine levels, and phasic dopamine release. Improved understanding of the role of dopamine-based reward circuitry and genotypes that influence the functioning of this circuitry may inform the design of more effective preventive and treatment interventions for obesity.

Introduction: Dopamine Reward and Obesity

Animal Models

Considerable research has implicated dopamine-based meso-limbic and meso-cortical circuitries in food reward. Extant data suggest that individuals with low levels of dopamine receptors are at increased risk for obesity. Compared to lean Zucker rats, obese Zucker rats, which have defective leptin receptor function, have fewer D2 receptors and reduced hypothalamic dopamine activity when fasting, but release more phasic dopamine when eating and do not stop eating in response to insulin and glucose administration [1]. In addition, D2 receptor blockade causes obese, but not lean, Zucker rats to overeat [2], implying that blockade of already low D2 receptor availability may sensitize obese rats to food. Obesity-prone (DIO) Sprague-Dawley rats, compared to the obesity-resistant strain, also have reduced dopamine turnover in the hypothalamus compared to the obesity-resistant strain before they become obese, and develop obesity only when given a palatable high-energy diet [3], both of which suggest a role for reward in the development of obesity. When exposed to the same high-fat diet, mice with lower D2 receptor density in the putamen show more weight gain than mice with higher D2 receptor density in this region [4].

Human Brain Imaging Studies

Functional brain imaging has become one of the most powerful tools available for analyses of the neural bases of human behavior. Here we emphasize its role in the analysis of dopamine-based reward. Two other chapters in this volume describe related aspects of brain imaging [5, 6].

Obese relative to lean humans show lower D2 receptor density in the striatum [7]. In interesting juxtaposition to these results, one PET study found that individuals who had recovered from anorexia nervosa, relative to healthy controls, had increased binding of D2/D3 receptors in the ventral striatum [8].

Brain imaging studies have found that obese versus lean individuals show reduced activation of the striatum, a key dopamine-target region, in response to food receipt [9, 10]. Conversely, women who have recovered from anorexia nervosa versus healthy controls showed greater activation of the striatum in response to a paradigm in which they receive financial reward [11]. Similarly, obese individuals show greater activation in the insula and anterior cingulate cortex in response to food receipt relative to healthy controls [10], whereas women who have recovered from anorexia nervosa show reduced responses in these areas in response to sweat tastes relative to controls [12].

As the studies reviewed above indicate, the relationship between altered dopamine activity and eating is not constant, with elevated dopamine activity sometimes associated with decreased eating and sometimes with increased eating. For example, a transgenic mouse with persistently elevated striatal dopamine displayed increased food intake and increased incentive motivation to consume food [13]. Furthermore, in patients with Parkinson's disease, treatment with dopamine agonists has been reported to occasionally cause compulsive eating and weight gain [14].

Genetic Variation in Dopaminergic Reward in Humans

Introduction

Given that dopamine plays a key role in food reward [15], it follows that genetic polymorphisms that affect dopamine neurotransmission may influence reward during or in anticipation of eating. Research has identified several genes that influence dopamine functioning, including those that affect dopamine receptors, transport, and breakdown.

The Taq1A Polymorphism of the DRD2 Gene

To date, the most empirical support has emerged for the *Taq1A* polymorphism of the DRD2 gene. The *Taq1A* polymorphism has three allelic variants: A1/A1, A1/A2, & A2/A2. *Taq1A* was originally thought to be located in the 3′-untranslated region of DRD2, but it actually resides in the neighboring ANKK1 gene [16]. Postmortem and positron emission tomography (PET) studies suggest that individuals with one or two copies of the A1 allele have 30–40% fewer D2 receptors and altered brain dopamine signaling compared to those without an A1 allele [17, 18]. Further data suggest that the A1 allele is associated with hypofunctioning of the prefrontal cortex, hypothalamus, striatum, insula, and amygdala [19, 20]. One hypothesis to explain these findings is that low D2 receptor density associated with the A1 allele makes individuals less sensitive to the activation of dopamine-based reward circuitry, rendering them more likely to overeat (and use psychoactive substances) to overcome this dopamine deficit [21].

Some [22, 23], but not all [24], studies found positive correlations between the A1 *TaqIA* allele and body mass index (BMI; weight in kg/height in m^2). Individuals with the A1 allele report greater food craving and work harder for food than those without this allele [25, 26]. Those with the A1 allele versus those without it are also more likely to abuse drugs and to experience greater cue-induced cravings for cigarettes and heroin [20, 27].

There is emerging evidence that the relation between abnormalities in food reinforcement and amount eaten is moderated by the A1 allele. Epstein et al. [25, 28] found an interaction between A1 allele and individual differences in food reward in adults, such that the greatest ad lib food intake occurred in those who reported more reinforcement from food and who had the A1 allele. Stice et al. [9] found that the relation between functional magnetic resonance imaging (fMRI) activation in response to consumption of palatable food (e.g., weaker caudate activation in response to milkshake receipt) was significantly more strongly related to current BMI and future weight gain over a 1-year follow-up in those with the A1 allele versus those without. Using another fMRI experimental paradigm, Stice et al. [unpubl. data] found that weaker activation of the frontal operculum, lateral orbitofrontal cortex, and striatum in response to imagined eating of appetizing foods, versus imagined eating of less palatable foods or drinking water, predicted elevated weight gain for those with the A1 allele. Collectively, these data suggest that reduced responsivity of eating-related dopaminergic reward circuitry increases the risk of obesity and that this relation is amplified in those who are thought to be at genetic risk for compromised dopamine signalling due to the presence of the A1 *TaqIA* DRD2 allele.

Variants in the DRD4 Gene

DRD4 variants have also been associated with reward sensitivity, impulsivity, and BMI. DRD4 is a postsynaptic dopamine receptor whose principal action is to inhibit the second messenger adenylate cyclase. D4 receptors are predominantly localized in areas that are innervated by mesocortical projections from the ventral tegmental area, including the prefrontal cortex, cingulate gyrus, and insula [29]. The 7-repeat or longer allele of this gene (DRD4-L) has been linked to compromised dopamine functioning in an in vitro study [30], to poorer response to dopamine-stimulating drugs [31], and to less dopamine release in the ventral caudate and nucleus accumbens after nicotine use [32], suggesting it may affect reward sensitivity. Presence of the DRD4-L allele has been associated with higher maximum BMI in humans at risk for obesity, including individuals with seasonal affective disorder who report overeating, individuals with bulimia nervosa [33, 34], and African-American adolescents [35]. Adults with versus without the DRD4-L allele have shown stronger food cravings in response to food cues [36] stronger smoking cravings and fMRI activation of the superior frontal gyrus and insula in response to smoking cues [37], stronger alcohol cravings in response to alcohol [38], and greater heroin craving in response to heroin cues [39]. Weaker activation of the frontal operculum in response to imagined eating of appetizing foods, versus imagined eating of less palatable foods or drinking water, predicted elevated weight gain for those with the 7-repeat or longer allele of the DRD4 gene [unpubl. data].

Variants in the Dopamine Transporter (DAT)

Phasically released dopamine is normally eliminated by rapid reuptake into presynaptic terminals through the dopamine transporter (DAT), which is abundant in the striatum [40]. Lower DAT expression, which is associated with a 10 repeat allele (DAT-L), may reduce synaptic clearance and produce higher basal, or tonic, dopamine levels and blunted phasic dopamine release [32]. Pecina et al. [13] found that a DAT knockdown mouse with increased extracellular dopamine displayed increased energy intake and preference for palatable foods. In a study of normal mice, feeding a high-fat diet significantly decreased DAT density in the dorsal and ventral parts of the caudate and putamen compared to a low-fat diet [41]. Lower striatal DAT levels also have been associated with elevated BMI in humans [42]. DAT-L has been associated with obesity in African-American smokers, versus smokers of other ethnic groups [43]. Finally, adults with versus without the DAT-L allele showed less dopamine release in response to cigarette smoking [32].

Variants in the Catechol-O-Methyltransferase Gene

Catechol-*O*-methyltransferase (COMT) also has been implicated in dopamine signaling. COMT regulates extrasynaptic dopamine breakdown, which occurs in both the prefrontal cortex and the striatum [44]. A single nucleotide exchange in the COMT gene, a valine to methionine substitution (Val/Met-158), is associated with a 4-fold reduction in COMT activity in humans. Individuals with the Met allele (low enzyme activity) had relatively higher tonic dopamine levels in prefrontal and striatal regions, but reduced phasic dopamine release in the striatum [45]. Indeed, individuals with versus without the Met-allele showed less phasic release of dopamine in response to cigarette smoking [32]. Although on study of patients with eating disorders found that individuals with the MET allele reported elevated bulimic symptoms than those without the MET allele [46], another study found that individuals with bulimia nervosa were marginally less likely to have a Met allele than were control individuals [47]. Interestingly, individuals with the Val/Val genotype also showed a preference for an immediate reward versus a larger, delayed reward [48], suggesting an important interplay between reward sensitivity and impulsivity.

AKT

AKT (also known as protein kinase B) is a protein kinase effector that acts downstream of the phosphatidylinositol 3-kinase (PI3K)-dependent intracellular signaling pathway. Overexpression of AKT decreases DAT expression and D2 receptor signaling [49]. Furthermore, the functioning of AKT is regulated by insulin; indeed, AKT appears to represent the primary pathway through which insulin affects dopamine transmission, including the amount of dopamine released during exposure to amphetamine [50]. This effect is mediated through changes in both the membrane-surface expression and the functionality of DAT. The importance of the AKT for amphetamine responsivity is demonstrated by the ability of the AKT1 inhibitor LY294002 to block amphetamine induced dopamine release and reuptake [50]. There is also evidence that phosphorylation of the AKT substrate PRAS40 is markedly reduced in rats fed a high-fat diet [51], suggesting a mechanism through which diet may influence DAT expression and dopamine release. Given this combination of clinical and preclinical data, we believe that AKT1 is an important candidate for understanding genetic relationships between obesity and dopamine functioning.

Dopamine and Learning

Emerging research has linked dopamine to learning. In a series of seminal experiments, Schultz [52] demonstrated that dopamine neuron firing occurs in response to unpredicted rewards or conditioned cues predictive of reward. It was soon realized that the firing pattern of dopamine neurons in these experiments resembled a theoretical learning signal that is referred to as a reward-prediction error signal in computational neuroscience [53]. Subsequently, studies using electrical recordings in animals and fMRI in humans have detected dopamine-related reward-prediction error signals in the brain [54]. These signals allow individuals to learn from positive and negative feedback and to shape future behavior to optimize the receipt of food and other biologically necessary rewards. Cohen and Frank [55] proposed that phasic dopamine bursts signal reward via dopamine D1 receptors, whereas pauses in dopamine firing signal omitted rewards via the D2 system. Recent evidence linking both the D1 and D2 systems to neuronal plasticity within the striatum supports this model [56]. If the model is correct, naturally occurring variations in levels of tonic and phasic dopamine, and of D1 and D2 receptors, could lead to differences in reward learning, and even promote aberrant reward processing, as seen in addiction and perhaps obesity. Consistent with this, two polymorphisms that lead to reduced expression of D2 receptors, *Taq-1A* A1 and *C957T*, also lead to impaired learning from negative feedback [57, 58]. It remains unclear how impaired negative-feedback learning could become a risk factor for addiction and obesity. Over-valuing rewarding aspects of certain behaviors while undervaluing their negative consequences, however, could be said to be a hallmark of addiction and over-eating. Behaviorally, this is referred to as an impulsive pattern, and indeed obese individuals tend to score higher on laboratory measures of impulsivity [59].

Dopamine is not only involved in learning about rewards, but also in their pursuit. As learning progresses, dopamine neuron firing occurs in response to conditioned cues predictive of reward [52], and this conditioned dopamine response appears to trigger approach and consumption of the associated reward [60]. The ability of cues to trigger appetite may be especially important in the current environment, in which we are constantly exposed to foods and food-related logos and advertising. Although extensive lesions of the dopamine system are associated with hypophagia [61], such animals are nonetheless able to learn about food rewards [62]. Their main deficit appears to be one of motivation, consistent with data showing that dopamine-deficient animals will not expend effort in order to obtain greater quantities of food. For example, animals with lesions of nucleus accumbens dopamine neurons given a choice between a high-effort, high-food density response versus a low-effort, low-food density response will typically choose the latter [63]. One might therefore predict, paradoxically, that low D2 receptors would be associated with an unwillingness to expend effort to obtain food. However, as stated earlier, low D2 receptors are associated with obesity in humans and animals. It is possible that our modern obesogenic environment, which provides abundant calories at low cost, overrides this tendency, if it is indeed present in humans.

Moreover, although considerable evidence links a heightened dopaminergic response to drug addiction and behavioral addictions such as pathological gambling

[64], reduced dopamine function has been associated with behavioral inflexibility in reward seeking and impulsivity [65]. In other words, deficient dopamine signaling could lead to persistent responding for rewards in the face of negative consequences, which could be the link between low dopamine function and obesity outlined above.

Conclusion

Paralleling processes that have been implicated in drug addiction, dopamine-based reward circuitry appears to play a role in encoding reward from eating and incentive sensitization, whereby cues associated with food reward acquire motivational value. Extant research suggests that low levels of dopamine D2 receptors and attenuated responsivity of dopamine-target regions (e.g. the striatum) to food and food cues are associated with elevated weight. There also is mixed evidence that genotypes that appear to lead to reduced dopamine signaling, including DRD2, DRD4 and DAT, are associated with obesity. In addition, there is emerging evidence that reduced responsivity of brain regions implicated in food reward increase risk for future weight gain in individuals who appear to be at genetic risk for attenuated dopamine signaling by virtue of DRD2 and DRD4 genotypes. It is crucial for future studies to critically re-examine these apparent relationships in independent prospective studies with increased statistical power. Furthermore, it is also important to use alternative brain imaging techniques, such as PET, to better characterize the specific parameters of dopaminergic synaptic function, such as dopamine receptor density, basal dopamine levels, and phasic dopamine release that mediate the effects of dopamine-related genotypes and responsivity of reward circuitry on risk for weight gain. Finally, future research should also investigate the role of dopamine-related genotypes on reward learning and on impulsivity, as these two factors may also increase risk for weight gain and interact with the responsivity of reward circuitry to natural rewards such as food. It is hoped that an improved understanding of the role of dopamine-based reward circuitry and genotypes that influence the functioning of this circuitry will inform the design of more effective preventive and treatment interventions for obesity.

References

1 Orosco, M, Rouch C, Nicolaidis S: Rostromedial hypothalamic monoamine changes in response to intravenous infusions of insulin and glucose in freely feeding obese Zucker rats: a microdialysis study. Appetite 1996;26:1–20.

2 Fetissov SO, Meguid MM, Sato T, Zhang LH: Expression of dopaminergic receptors in the hypothalamus of lean and obese Zucker rates and food intake. Am J Physiol Regul Integr Comp Physiol 2002;283:905–910.

3 Levin BE, Meynell AA: Defense of body weight depends on dietary composition and palatability in rats with diet-induced obesity. Am J Physiol Regul Integr Comp Physiol 2002;282:R46–R54.

4 Huang XF, Zavitsanou K, Huang X, Yu Y, Wang H, Chen F, Lawrence AJ, Deng C: Dopamine transporter and D2 receptor binding densities in mice prone or resistant to chronic high fat diet-incuded obesity. Behav Brain Res 2006;175:415–419.

5 Kringelbach ML, Stein A: Cortical mechanisms of human eating; in Langhans W, Greary N (eds): Frontiers in Eating and Weight Regulation. Forum Nutr. Basel, Karger, 2010, vol 63, pp 164–175.

6 Neary MT, Batterham RL: Gaining new insights into food reward with functional neuroimaging; in Langhans W, Greary N (eds): Frontiers in Eating and Weight Regulation. Forum Nutr. Basel, Karger, 2010, vol 63, pp 152–163.

7 Wang GJ, Volkow ND, Logan J, Pappas NR, Wong CT, Zhu W, Netusil N, Fowler JS: Brain dopamine and obesity. Lancet 2001;357:354–357.

8 Frank GK, Bailer UF, Henry SE, Drevets W, Meltzer CC, Price JC, Mathis CA, Wagner A, Hoge J, Ziolko S, Barbarich-Marsteller N, Weissfeld L, Kaye WH: Increased dopamine D2/D3 receptor binding after recovery from anorexia nervosa measured by positron emission tomography and [^{11}C]raclopride. Biol Psychiatry 2003;58:908–912.

9 Stice E, Spoor S, Bohon C, Small D: Relation between obesity and blunted striatal response to food is moderated by the TaqIA1 gene. Science 2008; 322:449–452.

10 Stice E, Spoor S, Bohon C, Veldhuizen M, Small D: Relation of reward from food intake and anticipated intake to obesity: a functional magnetic resonance imaging study. J Abnorm Psychol 2008;117:924–935.

11 Wagner A, Aizenstein H, Venkatraman VK, Fudge J, May JC, Mazurkewicz L, Frank GK, Bailer UF, Fischer L, Van N, Carter C, Putnam K, Kaye WH: Altered reward processing in women recovered from anorexia nervosa. Am J Psychol 2007;164:1842–1849.

12 Wagner A, Aizenstein H, Venkatraman VK, Fudge J, May JC, Mazurkewicz L, Frank GK, Bailer UF, Fischer L, Van N, Carter C, Putnam K, Kaye WH: Altered insula response to a taste stimulus in individuals recovered from restricting-type anorexia nervosa. Neuropscyhopharmacology 2008;33:513–523.

13 Pecina S, Cagniard B, Berridge KC, Aldrigde JW, Zhuang X: Hyperdopaminergic mutant mice have higher 'wanting' but not 'liking' for sweet rewards. J Neurosci 2003;23:9395–9402.

14 Nirenberg MJ, Waters C: Compulsive eating and weight gain related to dopamine agonist use. Mov Disord 2006;21:524–529.

15 Small DM, Jones-Gotman M, Dagher A: Feeding-induced dopamine release in dorsal striatum correlates with meal pleasantness ratings in healthy human volunteers. Neuroimage 2003;19:1709–1715.

16 Fossella J, Green AE, Fan J: Evaluation of a structural polymorphism in the ankyrin repeat and kinase domain containing 1 (ANKK1) gene and the activation of executive attention networks. Cogn Affect Behav Neurosci 2006;6:71–78.

17 Ritchie T, Noble EP: Association of seven polymorphisms of the D2 dopamine receptor gene with brain receptor-binding characteristics. Neurochem Res 2003;28:73–82.

18 Tupala E, Hall H, Bergström K, Mantere T, Räsäsen P, Särkioja R, Tiihonen J: Dopamine D2 receptors and transporters in type 1 and 2 alcoholics measured with human whole hemisphere autoradiography. Hum Brain Mapp 2003;20:91–102.

19 Bowirrat A, Oscar-Berman M: Relationship between dopaminergic neurotransmission, alcoholism, and reward deficiency syndrome. Am J Med Genet 2005; 132B:29–37.

20 Li Y, Shao C, Zhang D, Zhao M, Lin L, Yan P, Xie Y, Jiang K, Jin L: The effect of dopamine D2, D5 receptor and transporter (SLC6A3) polymorphisms on the cue-elicited heroin craving in Chinese. Am J Med Genet 2006;141B:269–273.

21 Volkow ND, Fowler JS, Wang GJ: Role of dopamine in drug reinforcement and addiction in humans: results from imaging studies. Behav Pharmacol 2002;13:355–66.

22 Spitz MR, Detry MA, Pillow P, Hu YH, Amos CI, Hong WK, Wu X: Variant alleles of the D2 dopamine receptor gene and obesity. Nutr Res 2000;20: 371–380.

23 Thomas GN, Critchley JA, Tomlinson B, Cockram CS, Chan JC: Relationships between the TaqI polymorphism of the dopamine D2 receptor and blood pressure in hyperglycemic and normoglycemic Chinese subjects. Clin Endocrinol (Oxf) 2001;55: 605–611.

24 Southon A. Walder K, Sanigorski AM, Zimmet P, Nicholson GC, Kotowicz MA, Collier G: The Taq IA and Ser311 Cys polymorphisms in the dopamine D2 receptor gene and obesity. Diabetes Nutr Metab 2003;16:72–76.

25 Epstein LJ, Leddy JJ, Temple JL, Faith MS: Food reinforcment and eating: a multilevel analysis. Psychol Bull 2007;133,884–906.

26 Comings DE, Flanagan SD, Dietz G, Muhleman D, Knell E, Gysin R: The dopamine D2 receptor (DRD2) as a major gene in obesity and height. Biochem Med Metab Biol 1993;50:176–185.

27 Erblich J, Lerman C, Self DW, Diaz GA, Bovbjerg DH: Effects of dopamine D2 receptor (DRD2) and transporter (SLC6A3) polymorphisms on smoking cue-induced cigarette craving among African American smokers. Mol Psychiatry 2005;10:407–414.

28 Epstein LH, Wright SM, Paluch RA, Leddy JJ, Hawk LW, Jaroni JL, Saad FG, Crystal-Mansour S, Shields PG, Lerman C: Relation between food reinforcement and dopamine genotypes and its effect on food intake in smokers. Am J Clin Nutr 2004;80:82–88.

29 Noain D, Avale ME, Wedemeyer C, Calvo D, Peper M, Rubinstein M: Identification of brain neurons expressin the dopamine D4 receptor gene using BAC transgenic mice. Eur J Neurosci 2006;24:2429–2438.

30 Asghari V, Sanyal S, Buchwaldt S, Paterson A, Jovanovic V, Van Tol HH: Modulation of intracellular cyclic AMP levels by different human dopamine D4 receptor variants. J Neurochem 1995;65:1157–1165.

31 Hamarman S, Fossella J, Ulger C, Brimacombe M, Dermody J: Dopamine receptor 4 (DRD4) 7-repeat allele predicts methylphenidate dose response in children with attention deficit hyperactivity disorders: a pharmacogentic study. J Child Adolesc Psychopharmacol 2004;14:564–574.

32 Brody AL, Mandelkern MA, Olmstead RE, Scheibal D, Hahn E, Shiraga S, Zamora-Paja E, Farahi J, Saxena S, London ED, McCracken JT: Gene variants of brain dopamine pathways and smoking-induced dopamine release in the ventral caudate/nucleus accumbens. Arch Gen Psychiatry 2006;63:808–816.

33 Kaplan AS, Levitan RD, Yilmaz Z, Davis C, Tharmalingam S, Kennedy JL: A DRD4/BDNF gene-gene interaction associated with maximum BMI in women with bulimia nervosa. Int J Eat Disord 2008;41:22–28.

34 Levitan RD, Masellis M, Lam RW, Muglia P, Basile VS, Jain U, Kaplan AS, Tharmakingam S, Kennedy SH, Kennedy JL: Childhood in-attention and dysphoria and adult obesity associated with the dopamine D4 receptor gene in overeating women with seasonal affective disorder. Neuropsychopharmacology 2004;29:179–86.

35 Guo G, North KE, Gorden-Larsen P, Bulik CM, Choi S: Body mass, DRD4, physical activity, sedentary behavior, and family socioeconomic status: The Add Health Study. Obesity 2007;15:1199–1206.

36 Sobik L, Hutchison K, Craighead L: Cue-elicited craving for food: a fresh approach to the study of binge eating. Appetite 2005;44:253–261.

37 McClernon FJ, Hutchison KE, Rose JE, Kozink RV: DRD4 VNTR polymorphism is associated with transient fMRI-BOLD responses to smoking cues. Psychopharmacology (Berl) 2007;194:433–441.

38 Hutchison KE, McGeary J, Smolen A, Bryan A, Swift RM: The DRD4 VNTR polymorphism moderates craving after alcohol consumption. Health Psychol 2002;21:139–146.

39 Shao C, Li Y, Jiang K, Zhang D, Xu Y, Lin L, Wang Q, Zhao M, Jin L: Dopamine D4 receptor polymorphism modulates cue-elicited heroin craving in Chinese. Psychopharmacology (Berl) 2006;186:185–190.

40 Floresco SB, West AR, Ash B, Moore H, Grace AA: Afferent modulation of dopamine neuron firing differentially regulates tonic and phasic dopamine transmission. Nat Neurosci 2003;6:968–973.

41 South T, Huang XF: High-fat diet exposure increases dopamine D2 receptor and decreases dopamine transporter receptor binding density in the nucleus accumbens and caudate putamen of mice. Neurochem Res 2008;33:598–605.

42 Chen PS, Yang YK, Yeh TL, Lee IH, Yao WJ, Chiu NT, Lu RB: Correlation between body mass index and striatal dopamine transporter availability in healthy volunteers: A SPECT study. Neuroimage 2008;40:275–279.

43 Epstein LH, Jaroni JL, Paluch RA, Leddy JJ, Vahue HE, Hawk L, Wileyto Pl, Shields PG, Lerman C: Dopamine transport genotype as a risk factor for obesity in smokers. Obesity 2002;10:1232–1240.

44 Matsumoto M, Weickert C, Beltaifa S, Kolachana B, Chen J, Hyde TM, Herman MM, Weinberger DR, Kleinman JE: Catechol-O-methyltransferase (COMT) mRNA expression in the dorsolateral prefrontal cortex of patients with schizophrenia. Neuropsychopharmacology 2003;28:1521–1530.

45 Bilder R, Volavka J, Lachman H, Grace A: The catechol-O-methyltransferase polymorphism: relations to the tonic-phasic dopamine hypothesis and neuropsychiatric phenotypes. Neuropsychopharmacology 2004;29:1943–1961.

46 Frieling H, Römer K, Konstanze D, Wilhelm J, Hillemacher T, Kornhuber J, de Zwaan M, Jacoby GE, Bleich S: Association of catecholamine-O-methyltransferase and 5-HTTLPR genotype with eating disorder-related behavior and attitudes in females with eating disorders. Psychiatr Genet 2006; 16:205–208.

47 Mikołajczyk E, Smiarowska M, Grzywacz A, Samochowiec J: Association of eating disorders with catechol-O-methyltransferase gene functional polymorphism. Neuropsychobiology 2006;54:82–86.

48 Boettiger CA, Mitchell JM, TavaresVC, Robertson M, Joslyn G, D'Esposito M, Fields HL: Immediate reward bias in humans: fronto-parietal networks and a role for the catechol-O-methyltransferase 158(Val/Val) genotype. J Neurosci 2007;27:14383–14391.

49 Garcia BG, Wei Y, Moron JA, Lin RZ, Javitch JA, Galli A: Akt is essential for insulin modulation of amphetamine-induced human dopamine transporter cell-surface redistribution. Mol Pharmacol 2005;68:102–109.

50 Williams JM, Owens WA, Turner GH, Saunders C, Dipace C, Blakely RD, France CP, Gore JC, Daws LC, Avison MJ, Galli A: Hypoinsulinemia regulates amphetamine-induced reverse transport of dopamine. PLoS Biol 2007;5:e274.
51 Nascimento EBM, Fodor M, van der Zon GCM, Jazet IM, Meinders AE, Voshol PJ, Vlasblom R, Baan B, Eckel J, Maassen JA, Diamant M, Ouwens DM: Insulin-mediated phosphorylation of the proline-rich AKT substrate PRAS40 is impaired in insulin target tissues of high-fat diet–fed rats. Diabetes 2007;55:3221–3228.
52 Schultz W: Predictive reward signal of dopamine neurons. J Neurophysiol 1998;80:1–27.
53 Montague PR, Dayan P, Sejnowski TJ: A framework for mesencephalic dopamine systems based on predictive Hebbian learning. J Neurosci 1996;16:1936–1947.
54 Schultz W: Behavioral theories and the neurophysiology of reward. Annu Rev Psychol 2006;57:87–115.
55 Cohen MX, Frank MJ: Neurocomputational models of basal ganglia function in learning, memory and choice. Behav Brain Res 2009;199:141–156.
56 Shen W, Flajolet M, Greengard P, Surmeier DJ: Dichotomous dopaminergic control of striatal synaptic plasticity. Science 2008;321:848–851.
57 Frank MJ, Moustafa AA, Haughey HM, Curran T, Hutchison KE: Genetic triple dissociation reveals multiple roles for dopamine in reinforcement learning. Proc Natl Acad Sci USA 2007;104:16311–16316.
58 Klein TA, Neumann J, Reuter M, Hennig J, von Cramon DY, Ullsperger M: Genetically determined differences in learning from errors. Science 2007;318:1642–1645.
59 Davis C: Psychobiological traits in the risk profile for overeating and weight gain. Int J Obes (Lond) 2009;33(suppl 2):49–53.
60 Phillips PE, Stuber GD, Heien ML, Wightman RM, Carelli RM: Subsecond dopamine release promotes cocaine seeking. Nature 2003;422:614–618.
61 Szczypka MS, Kwok K, Brot MD, Marck BT, Matsumoto AM, Donahue BA, Palmiter RD: Dopamine production in the caudate putamen restores feeding in dopamine-deficient mice. Neuron 2001;30:819–828.
62 Robinson S, Sandstrom SM, Denenberg VH, Palmiter RD: Distinguishing whether dopamine regulates liking, wanting, and/or learning about rewards. Behav Neurosci 2005;19:5–15.
63 Cousins MS, Atherton A, Turner L, Salamone JD: Nucleus accumbens dopamine depletions alter relative response allocation in a T-maze cost/benefit task. Behav Brain Res 1996;74:189–97.
64 Dagher A, Robbins TW: Personality, addiction, dopamine: insights from Parkinson's disease. Neuron 2009;61:502–510.
65 Cardinal RN, Pennicott DR, Sugathapala CL, Robbins TW, Everitt BJ: Impulsive choice induced in rats by lesions of the nucleus accumbens core. Science 2001;292:2499–2501.

Eric Stice, PhD
Oregon Research Institute
1715 Franklin Blvd.
Eugene, OR 97403 (USA)
Tel. +1 541 484 2123, Fax +1 541 484 1108, E-Mail estice@ori.org

Langhans W, Geary N (eds): Frontiers in Eating and Weight Regulation.
Forum Nutr. Basel, Karger, 2010, vol 63, pp 186–194

Metabolic Imprinting in Obesity

E.L. Sullivan · K.L. Grove

Division of Neuroscience, Oregon National Primate Research Center (ONPRC), Oregon Health and Science University, Beaverton, Oreg., USA

Abstract

Increasing evidence indicates that early metabolic programming contributes to escalating obesity rates in children and adults. Metabolic imprinting is involved in the establishment of set points for physiologic and metabolic responses in adulthood. Evidence from epidemiological studies and animal models indicates that maternal health and nutritional status during gestation and lactation have long-term effects on central and peripheral systems that regulate energy balance in the developing offspring. Perinatal nutrition also impacts susceptibility to developing metabolic disorders and plays a role in programming body weight set points. The states of maternal energy status and health that are implicated in predisposing offspring to increased risk of developing obesity include maternal overnutrition, diabetes, and undernutrition. This chapter discusses the evidence from epidemiologic studies and animal models that each of these states of maternal energy status results in metabolic imprinting of obesity in offspring. Also, the potential molecular mediators of metabolic imprinting of obesity by maternal energy status including glucose, insulin, leptin, inflammatory cytokines and epigenetic mechanisms are considered.

Introduction

Obesity Epidemic

The increasing prevalence of obesity has large implications for the health of the human population [1]. The escalating incidence of obesity is due not only to environmental changes such as increased availability of energy dense food and decreased physical exertion, but mounting evidence in humans and animal models indicates that early programming events also play an important role. Maternal health and nutritional status during gestation and lactation have long-term effects on central and peripheral systems that regulate energy balance in offspring.

Metabolic Imprinting

Individuals best suited to their environment are most likely to reproduce and pass on their genes. A critical factor affecting reproductive success is the type and availability of food sources. Survival is optimized if energy balance regulation is programmed to most efficiently use available metabolic fuels. Metabolic imprinting is the process by which a stimulus or insult during a critical period of development has a long-term effect on the metabolic responses of the offspring. During development, mammals are exposed to two environments: the in utero and postnatal environments. Maternal diet, body composition, and energy stores have major influences on both environments. Perinatal nutrition also has long lasting effects on energy balance regulation, influences susceptibility to developing metabolic disorders [2, 3] and plays a role in programming body weight set points [4, 5]. The states of maternal energy status and health implicated in predisposing offspring to obesity include maternal overnutrition, diabetes, and undernutrition. The current research on each of these states is discussed in the following sections. Because it is not within the scope of this chapter to cover all of the literature, representative and seminal pieces of work are highlighted.

Maternal Undernutrition

Epidemiological Studies

The consequences of maternal undernutrition on offspring energy balance regulation were first examined in humans who experienced the Dutch famine during gestation. Ravelli and coworkers [6, 7] discovered that famine during different periods of gestation led to differential risks of various metabolic disorders. The idea that undernourishment during gestation affects later risk for obesity and metabolic disorders is supported by the findings of Barker et al. [8] who reported associations between low birth weight, an indirect measure of the fetal environment, and increased risk of cardiovascular disease, stroke, hypertension, and diabetes mellitus [9]. These findings led to the 'fetal origins hypothesis' which postulates that the body's structure, physiology and metabolism are programmed during embryonic and fetal life [9], and the 'thrifty phenotype hypothesis', which suggests that undernourishment during development causes an adaptive response that programs offspring to prioritize organ growth and increases metabolic efficiency in preparation for an environment with sparse resources [10]. Such programming becomes detrimental, however, when postnatal nutrition is plentiful and offspring exhibit rapid catch up growth and obesity [11].

Animal Models of Maternal Undernutrition

Prenatal undernutrition has been extensively modeled in rodents by restricting the amount of calories received by the mother during different stages of gestation. Results vary depending on the timing, length and magnitude of the food restriction. For example, Jones and colleagues [12, 13] found that restricting food access by 50% during the first two trimesters of pregnancy produced hyperphagia and delayed-onset obesity when maintained on either a control or a high-fat diet (HFD) in male (but not female) offspring. Offspring that experienced severe 70% dietary restriction throughout gestation exhibited adult-onset obesity, hyperinsulinemia, and hyperleptinemia associated with hyperphagia and hypoactivity [2, 14]. However, even a mild 30% food restriction during pregnancy led to reduced birth weight, catch-up growth and hypersensitivity to HFD-induced obesity [5]. While the mechanisms leading to long-term reprogramming are poorly understood, leptin has been identified as a key factor. Undernourished offspring exhibit premature onset of the neonatal leptin surge [5], and when the leptin surge was artificially induced prematurely in control offspring, they also exhibited hypersensitivity to HFD-induced obesity. Therefore, both the magnitude and timing of the leptin surge appear to be important for the development of metabolic systems and their neuronal controls. This is further discussed by Bouret [15] in this volume.

Rodent models have many advantages, such as short gestation and the ability to manipulate genetics. However, the critical periods for development of energy balance regulatory systems differ between rodents and humans. In rodents, the neural pathways regulating energy balance are immature at birth and are not fully developed until the third postnatal week (mice) [16]. In contrast, in humans, nonhuman primates (NHP), pigs, and sheep, hypothalamic eating-control circuits develop primarily prenatally [17]. Thus, models of maternal undernutrition in which the development of energy balance regulation occurs prenatally are particularly relevant. Nathanielsz and colleagues developed a NHP model of maternal undernutrition in which pregnant females are 30% restricted from early to mid-gestation [18]. In this model, undernourished mothers have fetal offspring with normal body weight, but decreased hip circumference and perturbations in kidney and liver development [18]. Future studies examining the phenotype of juvenile and adult offspring are critical to understanding the effects of maternal undernutrition on obesity susceptibility in primates.

Maternal Overnutrition

Epidemiological Studies

Because most humans in developed countries experience an environment in which food scarcity is rare and energy dense foods are readily available, the most common perturbation of maternal nutritional status in developed countries is maternal obesity and overnutrition. Currently, over one third of pregnant American women are obese [19], and most of them consume an excess amount of food and fat [20]. Epidemiological studies show that maternal obesity increases the incidence of obesity and metabolic syndrome in children [21]. The effect of maternal obesity on the susceptibility to obesity in offspring is independent of gestational diabetes, as obese women with normal blood glucose have babies with increased adiposity [22]. Several animal models are used to study the effects of overnutrition during gestation and the early postnatal period on the developing offspring.

Animal Models of Maternal Overnutrition

Promoting obesity by feeding animals a HFD is a common model of maternal overnutrition. For example, pups from rats fed a HFD during pregnancy and lactation were heavier, fatter, hyperglycemic and had higher hepatic lipid content at weaning than pups from control mothers [23]. Similarly, in a mouse model of chronic maternal overnutrition, offspring were hyperphagic, and had reduced locomotion and increased adiposity at 3 and 6 months of age [24]. Another group demonstrated that offspring of rats fed a highly palatable processed junk-food or cafeteria diet during gestation and lactation became heavier and displayed an increased preference for fatty, sugary and salty foods [25]. Although in the majority of studies offspring from HFD-fed mothers were overweight, several studies reported that maternal HFD consumption produced lighter offspring, potentially due to impaired lactation in obese mothers [26]. Differences in the duration of HFD consumption (i.e. chronically or only during gestation and lactation) and fatty acid composition of the diets are hypothesized to explain differences in offspring phenotype [26].

Rodents genetically predisposed to obesity are also used to examine the effects of maternal obesity. In a study of the obese agouti mouse [27], wild-type offspring of agouti dams bred to wild-type males were heavier than offspring from wild-type crosses. Interestingly, no difference in adult weight was found. Heterozygous leptin receptor-deficient mice ($lepr^{db/+}$) are used to model maternal obesity because they overeat and have increased weight gain during pregnancy [28]. In this model, offspring of $lep^{db/+}$ females are heavier than controls regardless of genotype. However, this model is complicated by the fact that the mothers also develop spontaneous gestational diabetes. Levin and colleagues [29, 30], examined the interaction between genetics and maternal overnutrition in substrains of Sprague-Dawley rats that were either resistant or sensitive to diet-induced obesity. Diet-sensitive rats that consumed a high-energy diet prior to and during pregnancy and lactation had offspring with increased adiposity and hyperglycemia when maintained on a control diet and increased weight gain and leptin levels when consuming a high-energy diet as compared to offspring of diet-sensitive mothers that consumed the control diet. In contrast, maternal diet did not affect weight gain in offspring of diet-resistant mothers, indicating that maternal obesity enhances obesity susceptibility in offspring with a genetic predisposition for diet-induced obesity [29].

The consequences of maternal obesity and HFD consumption on offspring energy balance regulation are being examined in a NHP model by our group. Surprisingly, fetal offspring of Japanese macaques fed a HFD diet, whether obese with severe insulin resistance or lean with normal insulin sensitivity, display signs of severe lipotoxicity. Juvenile offspring (6 months old) from HFD-fed mothers were heavier and have increased adiposity, leptin levels and show signs of fatty liver disease [31]. This suggests that in NHP, as in rodents, maternal overnutrition predisposes offspring to early-onset obesity and metabolic disorders.

The effects of early postnatal overnutrition have been examined in a variety of animal models. Young adult female baboons overfed as infants were heavier and had increased adiposity compared to control-fed females [32]. Surprisingly, however, overfeeding prior to weaning increased adiposity in male baboons, but did not affect body weight [32]. Postnatal overfeeding is commonly examined in rodents by adjusting the number of pups per litter; pups raised in small litters receive increased nutrition and pups raised in large litters receive reduced nutrition. This paradigm has led to a number of interesting findings: overfeeding during suckling has long-term effects on energy balance regulation, as adult rats raised in small litters have increased body weight, adiposity [33], leptin resistance [34] and abnormalities in the sensitivity of hypothalamic neurons to several neuropeptide and nutrient signals [35]. Our group has shown that overfed offspring are hypersensitive to a HFD in adulthood leading to accelerated weight gain and metabolic disorders, apparently due at least in part to long-term defects in hypothalamic leptin sensitivity [36].

Maternal Diabetes

Epidemiological Studies

The insulin and leptin resistances associated with pregnancy increase women's susceptibility to developing gestational diabetes, which occurs in 3–14% of pregnancies in the USA [17]. Mothers with gestational diabetes have babies with increased birth weight and increased risk of developing childhood obesity and type 2 diabetes mellitus (T2DM) [37]. The type and severity of maternal diabetes influences the offspring's outcome. Obesity and impaired glucose tolerance are 2–3 times more frequent in offspring of mothers with T2DM than offspring of mothers who developed gestational diabetes [38]. Maternal diabetes (either type 1 diabetes mellitus or T2DM) affects offspring during the postnatal period, as mothers with diabetes that persists after gestation have breast milk with increased glucose and insulin levels [39]. Plagemann et al. [39] found that infants breast-fed by diabetic mothers had a greater risk of developing obesity and impaired glucose metabolism than infants from diabetic mothers fed breast milk from nondiabetic women. This finding has important clinical implications and needs to be examined further so that diabetic mothers can be advised on the optimal source of nutrition for their infants.

Animal Models of Maternal Diabetes

Animal models of maternal diabetes provide strong evidence that maternal diabetes increases susceptibility to obesity and diabetes [28, 40]. Female rats made hyperglycemic during pregnancy by continuous glucose infusions produced offspring with glucose intolerance and impaired insulin secretion [41]. Also, hyperglycemia induced by the pancreatic islet toxin streptozotocin during early pregnancy produced offspring that were macrosomic at birth and had elevated weight gain during the first ten weeks of life [42]. Studies using genetic models of spontaneous diabetes, such as the $lepr^{db/+}$ mouse described above, find that offspring of mothers with gestational diabetes have increased adiposity and insulin resistance [28].

Molecular Mechanisms Underlying Metabolic Imprinting

It is important to understand the mechanisms by which metabolic imprinting leads to particular phenotypes and influences susceptibility to obesity and metabolic diseases. The mediators and pathways that transmit signals from the mother to program the metabolic phenotype of the developing offspring are not fully elucidated. Hormones such as leptin and insulin, nutrients such as glucose, free fatty acids, and triglycerides and inflammatory cytokines are implicated.

Maternal glucose crosses the placenta and is transferred to the fetus. However, as maternal insulin cannot cross the placenta [21], the fetal pancreas must secrete insulin in response to maternal glucose. Maternal overnutrition and diabetes produce maternal hyperglycemia, which increases fetal insulin secretion [43]. Fetal hyperinsulinemia, in turn, is hypothesized to be involved in the programming of diabetes and obesity [44]. This is supported by studies showing that insulin administration to rats during the third trimester of pregnancy produces obese offspring [3, 13]. Also, intrahypothalamic administration of insulin to rat pups during the development of projections from the arcuate (Arc) to the paraventricular (PVH) nuclei caused increased body weight, hyperinsulinemia, impaired glucose tolerance, and increased susceptibility to diabetes [45]. As insulin is a neuronal growth factor [44], fetal hyperinsulinemia may cause perturbations in the development of energy balance regulatory circuitry, increasing susceptibility to obesity and metabolic diseases.

Leptin is also implicated in programming obesity. Studies in mice show that postnatal leptin plays an important role in the development of hypothalamic neuronal connections [16] (see also [15] in this volume). In humans, leptin is increased in maternal obesity and diabetes [46, 47] and reduced in babies with intrauterine growth restriction [48]. However, leptin levels do not increase until the end of the third trimester of gestation [48], after hypothalamic development. Fetal leptin levels are low to undetectable during important developmental stages in NHP and do not increase until late in the third trimester [49]. Thus, while it is clear that leptin is an important neurotrophic factor in rodents, its role in NHP and humans remains unclear.

Inflammatory cytokines are elevated in obese pregnant women [50] and have been postulated to be a potential mediator of metabolic imprinting. Using a NHP model, we have reported that fetuses from mothers consuming a HFD had elevated circulating and hypothalamic cytokines [unpubl. obs.]. Rodent studies show that both agouti-related peptide [51] and pro-opiomelanocortin [52] neurons in the Arc are directly affected by cytokines, leading Marks and colleagues to hypothesize that fetal exposure to increased cytokines directly impacts the development of neurons regulating energy balance.

Epigenetic mechanisms of imprinting individual differences in obesity susceptibility are receiving increasing attention. Epigenetics is the study of changes in gene expression not involving changes in DNA sequence, such as DNA methylation, covalent histone modifications, packaging of DNA around nucleosomes, folding of chromatin and attachment of chromatin to the nuclear matrix [53]. Epigenetic mechanisms are affected by perinatal nutrition, maternal energy status, maternal endocrine status and oxidative stress [53]. For example, maternal calorie restriction is associated with increased methylation of the Hras gene, and maternal protein deficiency is associated with alterations in epigenetic regulation of the glucocorticoid receptor and peroxisome proliferator activated receptor alpha. Also, in utero exposure to deficiency/excess of various nutrients causes changes in epigenetic modifications [53]. Such data indicate that it is likely that maternal energy status influences susceptibility to obesity in the offspring by causing perturbations in epigenetic modifications. However, further research is needed to determine the effects of maternal overnutrition, undernutrition and diabetes on epigenetic modifications and to understand the mechanisms underlying these changes.

References

1 Singhal V, Schwenk WF, Kumar S, Evaluation and management of childhood and adolescent obesity. Mayo Clin Proc 2007;82:1258–1264.

2 Vickers MH, Breier BH, Cutfield WS, Hofman PL, Gluckman PD: Fetal origins of hyperphagia, obesity, and hypertension and postnatal amplification by hypercaloric nutrition. Am J Physiol Endocrinol Metab 2000;279:E83–E87.

3 Jones AP, Dayries M: Maternal hormone manipulations and the development of obesity in rats. Physiol Behav 1990;47:1107–1110.

4 Vickers MH, Gluckman PD, Coveny AH, Hofman PL, Cutfield WS, Gertler A, Breier BH, Harris M: Neonatal leptin treatment reverses developmental programming. Endocrinology 2005;146:4211–4216.

5 Yura S, Itoh H, Sagawa N, Yamamoto H, Masuzaki H, Nakao K, Kawamura M, Takemura M, Kakui K, Ogawa Y, Fujii S: Role of premature leptin surge in obesity resulting from intrauterine undernutrition. Cell Metab 2005;1:371–378.
6 Ravelli AC, van Der Meulen JH, Osmond C, Barker DJ, Bleker OP: Obesity at the age of 50 y in men and women exposed to famine prenatally. Am J Clin Nutr 1999;70:811–816.
7 Roseboom TJ, van der Meulen JH, Ravelli AC, Osmond C, Barker DJ, Bleker OP: Effects of prenatal exposure to the Dutch famine on adult disease in later life: an overview. Mol Cell Endocrinol 2001;185:93–98.
8 Barker DJ, Winter PD, Osmond C, Margetts B, Simmonds SJ: Weight in infancy and death from ischaemic heart disease. Lancet 1989;ii:577–580.
9 Barker DJ: The fetal and infant origins of disease. Eur J Clin Invest 1995;25:457–463.
10 Hales CN, Barker DJ: Type 2 (non-insulin-dependent) diabetes mellitus: the thrifty phenotype hypothesis. Diabetologia 1992;35:595–601.
11 Barker DJ, Eriksson JG, Forsen T, Osmond C: Fetal origins of adult disease: strength of effects and biological basis. Int J Epidemiol 2002;31:1235–1239.
12 Jones AP, Friedman MI: Obesity and adipocyte abnormalities in offspring of rats undernourished during pregnancy. Science 1982;215:1518–1519.
13 Jones AP, Pothos EN, Rada P, Olster DH, Hoebel BG: Maternal hormonal manipulations in rats cause obesity and increase medial hypothalamic norepinephrine release in male offspring. Brain Res Dev Brain Res 1995;88:127–131.
14 Vickers MH, Breier BH, McCarthy D, Gluckman PD: Sedentary behavior during postnatal life is determined by the prenatal environment and exacerbated by postnatal hypercaloric nutrition. Am J Physiol Regul Integr Comp Physiol 2003;285:R271–R273.
15 Bouret SG: Development of hypothalamic neural networks controlling appetite; in Langhans W, Greary N (eds): Frontiers in Eating and Weight Regulation. Forum Nutr. Basel, Karger, 2010, vol 63, pp 84–93.
16 Djiane J, Attig L: Role of leptin during perinatal metabolic programming and obesity. J Physiol Pharmacol 2008;59(suppl 1):55–63.
17 Grove KL, Grayson BE, Glavas MM, Xiao XQ, Smith MS: Development of metabolic systems. Physiol Behav 2005;86:646–660.
18 Li C, Schlabritz-Loutsevitch NE, Hubbard GB, Han V, Nygard K, Cox LA, McDonald TJ, Nathanielsz PW: Effects of maternal global nutrient restriction on fetal baboon hepatic IGF system genes and gene products. Endocrinology 2009;4634–4642.
19 King JC: Maternal obesity, metabolism, and pregnancy outcomes. Annu Rev Nutr 2006;26:271–291.
20 Alberti-Fidanza A, Parizkova J, Fruttini D: Relationship between mothers' and newborns' nutritional and blood lipid variables. Eur J Clin Nutr 1995;49:289–298.
21 Oken E, Gillman MW: Fetal origins of obesity. Obes Res 2003;11:496–506.
22 Sewell MF, Huston-Presley L, Super DM, Catalano P: Increased neonatal fat mass, not lean body mass, is associated with maternal obesity. Am J Obstet Gynecol 2006;1100–1103.
23 Guo F, Jen KL: High-fat feeding during pregnancy and lactation affects offspring metabolism in rats. Physiol Behav 1995;57:681–686.
24 Samuelsson AM, Matthews PA, Argenton M, Christie MR, McConnell JM, Jansen EH, Piersma AH, Ozanne SE, Twinn DF, Remacle C, Rowlerson A, Poston L, Taylor PD: Diet-induced obesity in female mice leads to offspring hyperphagia, adiposity, hypertension, and insulin resistance: a novel murine model of developmental programming. Hypertension 2008;51:383–392.
25 Bayol SA, Farrington SJ, Stickland NC: A maternal 'junk food' diet in pregnancy and lactation promotes an exacerbated taste for 'junk food' and a greater propensity for obesity in rat offspring. Br J Nutr 2007;98:843–851.
26 Ferezou-Viala J, Roy AF, Serougne C, Gripois D, Parquet M, Bailleux V, Gertler A, Delplanque B, Djiane J, Riottot M, Taouis M: Long-term consequences of maternal high-fat feeding on hypothalamic leptin sensitivity and diet-induced obesity in the offspring. Am J Physiol Regul Integr Comp Physiol 2007;293:R1056–R1062.
27 Han J, Xu J, Epstein PN, Liu YQ: Long-term effect of maternal obesity on pancreatic beta cells of offspring: reduced beta cell adaptation to high glucose and high-fat diet challenges in adult female mouse offspring. Diabetologia 2005;48:1810–1818.
28 Yamashita H, Shao J, Qiao L, Pagliassotti M, Friedman JE: Effect of spontaneous gestational diabetes on fetal and postnatal hepatic insulin resistance in Lepr(db/+) mice. Pediatr Res 2003;53:411–418.
29 Levin BE, Govek E: Gestational obesity accentuates obesity in obesity-prone progeny. Am J Physiol 1998;275:R1374–R1379.
30 Levin BE: Metabolic imprinting on genetically predisposed neural circuits perpetuates obesity. Nutrition 2000;16:909–915.
31 McCurdy CE, Bishop JM, Williams SM, Grayson BE, Smith MS, Friedman JE, Grove KL: Maternal high-fat diet triggers lipotoxicity in the fetal livers of nonhuman primates. J Clin Invest 2009;119:323–335.

32 Lewis DS, Bertrand HA, McMahan CA, McGill HC Jr, Carey KD, Masoro EJ: Preweaning food intake influences the adiposity of young adult baboons. J Clin Invest 1986;78:899–905.
33 Levin BE: Metabolic imprinting: critical impact of the perinatal environment on the regulation of energy homeostasis. Philos Trans R Soc Lond B Biol Sci 2006;361:1107–1121.
34 Schmidt I, Schoelch C, Ziska T, Schneider D, Simon E, Plagemann A: Interaction of genetic and environmental programming of the leptin system and of obesity disposition. Physiol Genomics 2000;3:113–120.
35 Bouret SG: Early life origins of obesity: role of hypothalamic programming. J Pediatr Gastroenterol Nutr 2009;48(suppl 1) S31–S38.
36 Glavas MM, Kirigiti MA, Xiao QX, Enriori PJ, Fisher SK, Evans AE, Grayson BE, Cowley MA, Smith MS, Grove KL: Early overnutrition results in early onset arcuate leptin resistance and increased sensitivity to high-fat diet. Diabetes 2009;unpublished observation.
37 Gillman MW, Rifas-Shiman S, Berkey CS, Field AE, Colditz GA: Maternal gestational diabetes, birth weight, and adolescent obesity. Pediatrics 2003;111: e221–e226.
38 Pettitt DJ, Nelson RG, Saad MF, Bennett PH, Knowler WC: Diabetes and obesity in the offspring of Pima Indian women with diabetes during pregnancy. Diabetes Care 1993;16:310–314.
39 Plagemann A, Harder T, Franke K, Kohlhoff R: Long-term impact of neonatal breast-feeding on body weight and glucose tolerance in children of diabetic mothers. Diabetes Care 2002;25:16–22.
40 Gauguier D, Nelson I, Bernard C, Parent V, Marsac C, Cohen D, Froguel P: Higher maternal than paternal inheritance of diabetes in GK rats. Diabetes 1994;43:220–224.
41 Gauguier D, Bihoreau MT, Ktorza A, Berthault MF, Picon L: Inheritance of diabetes mellitus as consequence of gestational hyperglycemia in rats. Diabetes 1990;39:734–739.
42 Oh W, Gelardi NL, Cha CJ: Maternal hyperglycemia in pregnant rats: its effect on growth and carbohydrate metabolism in the offspring. Metabolism 1988;37:1146–1151.
43 Leung TW, Lao TT: Placental size and large-for-gestational-age infants in women with abnormal glucose tolerance in pregnancy. Diabet Med 2000;17: 48–52.
44 Simerly RB: Hypothalamic substrates of metabolic imprinting. Physiol Behav 2008;94:79–89.
45 Plagemann A, Heidrich I, Gotz F, Rohde W, Dorner G: Lifelong enhanced diabetes susceptibility and obesity after temporary intrahypothalamic hyperinsulinism during brain organization. Exp Clin Endocrinol 1992;99:91–95.
46 Lepercq J, Hauguel-De Mouzon S, Timsit J, Catalano PM: Fetal macrosomia and maternal weight gain during pregnancy. Diabetes Metab 2002;28:323–328.
47 Hauguel-de Mouzon S, Shafrir E: Carbohydrate and fat metabolism and related hormonal regulation in normal and diabetic placenta. Placenta 2001;22: 619–627.
48 Davidowa H, Plagemann A: Decreased inhibition by leptin of hypothalamic arcuate neurons in neonatally overfed young rats. Neuroreport 2000;11: 2795–2798.
49 Grayson BE, Allen SE, Billes SK, Williams SM, Smith MS, Grove KL: Prenatal development of hypothalamic neuropeptide systems in the nonhuman primate. Neuroscience 2006;143:975–986.
50 Stewart FM, Freeman DJ, Ramsay JE, Greer IA, Caslake M, Ferrell WR: Longitudinal assessment of maternal endothelial function and markers of inflammation and placental function throughout pregnancy in lean and obese mothers. J Clin Endocrinol Metab 2007;92:969–975.
51 Scarlett JM, Zhu X, Enriori PJ, Bowe DD, Batra AK, Levasseur PR, Grant WF, Meguid MM, Cowley MA, Marks DL: Regulation of agouti-related protein messenger ribonucleic acid transcription and peptide secretion by acute and chronic inflammation. Endocrinology 2008;149:4837–4845.
52 Scarlett JM, Jobst EE, Enriori PJ, Bowe DD, Batra AK, Grant WF, Cowley MA, Marks DL: Regulation of central melanocortin signaling by interleukin-1 beta. Endocrinology 2007;148:4217–4225.
53 Campion J, Milagro FI, Martinez JA: Individuality and epigenetics in obesity. Obes Rev 2009;383–392.

Kevin L. Grove
ONPRC
505 NW 185th
Beaverton, OR 97006 (USA)
Tel. +1 503 690 5380, Fax +1 503 690 5384, E-Mail grovek@ohsu.edu

Langhans W, Geary N (eds): Frontiers in Eating and Weight Regulation.
Forum Nutr. Basel, Karger, 2010, vol 63, pp 195–203

Gene-Environment Interactions in Obesity

Marion M. Hetherington[a] · Joanne E. Cecil[b]

[a]Institute of Psychological Sciences, University of Leeds, Leeds, and [b]Bute Medical School, University of St Andrews, St Andrews, UK

Abstract

Obesity is a global and growing problem. The detrimental health consequences of obesity are significant and include co-morbidities such as diabetes, cancer and coronary heart disease. The marked rise in obesity observed over the last three decades suggests that behavioural and environmental factors underpin the chronic mismatch between energy intake and energy expenditure. However, not all individuals become obese, suggesting that there is considerable variation in responsiveness to 'obesogenic' environments. Some individuals defend easily against a propensity to accumulate fat mass and become overweight whilst others are predisposed to gain weight, possibly as a function of genotype. The genetic contribution to obesity is well established. Common obesity is polygenic, involving complex gene-gene and gene-environment interactions, and it is these interactions that produce the multi-factorial obese phenotypes. Candidate gene variants for polygenic obesity appear to disrupt pathways involved in the regulation of energy intake and expenditure and include adrenergic receptors, uncoupling proteins, PPARG, POMC, MC4R and a set of single nucleotide polymorphisms in the FTO locus. Notably, the FTO gene is the most robust gene for common obesity characterised to date, and recent data shows that the FTO locus seems to confer risk of obesity through increasing energy intake and reduced satiety. Gene variants involved in pathways regulating addiction and reward behaviours may also play a role in predisposition to obesity. Understanding the routes through which the genotype is expressed will ultimately provide opportunities for developing strategies to intervene, as the interaction between genotype and environment is potentially modifiable through behaviour change.

Overview – The Case for the Environment

Obesity is a state of excess adiposity. Accurate measurement of adiposity is generally eschewed in large populations; rather, body mass index (BMI: wt/ht^2) is used as a proxy measure, with a BMI of 29.9 at the cut-off for adult obesity. The prevalence of obesity is increasing globally and is no longer simply associated with industrialised, developed nations. One third of US adults and 1 in 4 UK adults has a BMI of 30 and above. Childhood obesity is also increasing and it is estimated that 1 in 10 of the world's children carry excess body fat [1].

While industrialised countries have a social gradient, with obesity prevalence highest among the least affluent, in many developing countries, obesity is still associated with affluence and status. For example, the prevalence of obesity in the Western Pacific nation Nauru is above 75% and that for women in Tonga is 70%. Factors which have influenced changes in obesity prevalence in this region include a long and continuing transition from a traditional, island diet of freshly caught fish, products from coconut and pandanus trees to a modern Western diet, 80% of which is imported, as islanders have become more affluent [2].

The detrimental health consequences of overweight and obesity are well documented and include increased risk of diabetes, certain cancers, coronary heart disease and hypertension [3]. Clearly, although fat mass plays a crucial role in maintaining energy stores and in providing a buffer against food shortage, any adaptive and protective benefit of obesity seems to have been thwarted by environmental pressures facilitating excess body fatness. Given the rapidity with which obesity has increased over the past 3 decades and the spread of this increase across the globe, it is clear that behavioural and environmental changes are fundamental to the obesity 'epidemic' [3]. The example of the Western Pacific peoples demonstrates this proposition. These nations have undergone rapid, dramatic and profound changes in the cultural, economic and social landscape since the first European explorers arrived in the 16th century and colonialists in the 18th century. Whilst there are no data of overweight prevalence from this time, early descriptions of the indigenous Pacific inhabitants were of a 'tall, muscular and well-proportioned people' [4, p. 13]. However, four hundred years of modernisation have taken their toll, including dependency by some Western Pacific nations on developed countries for food [2]. Given the relatively long genetic isolation of Pacific islanders [5] and the relatively short time it has taken for obesity levels to increase, it is clear that behavioural and environmental change has been necessary for the increase in obesity. Modernisation of the diet from traditional, fresh and local to pre-prepared, processed and imported as well as fewer opportunities for energy expenditure through labour-saving devices contribute to the same energy imbalance as experienced elsewhere in the world. Thus, obesity is caused by an excess of intake relative to energy expenditure sustained over time regardless of age, or whether the individual is from a developed or developing nation.

Gene-Environment Interactions

At the population level, obesity is commonplace under facilitative environmental conditions. The impact of weight-promoting environments is best characterised by the experience of Pima Indians living in the Sierra Madre mountains of Northern Mexico and their westernized Pima counterparts living mainly in reservations in Arizona. Mexican Pima have relatively low levels of obesity (7% men; 20% women) and type 2 diabetes (T2DM) (7%) comparable with non-Pima Mexicans; Pima Indians in

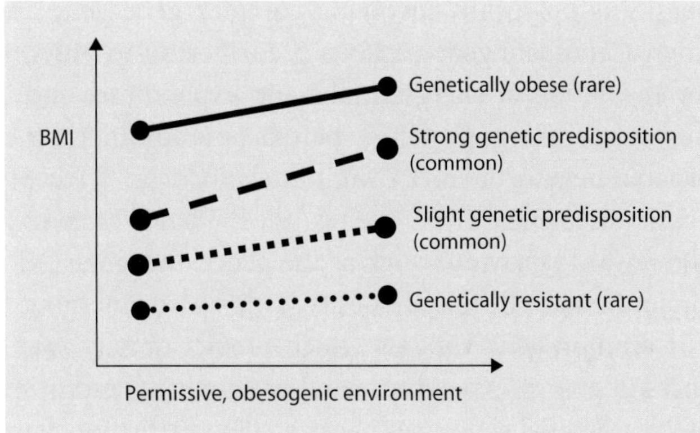

Fig. 1. Interaction between genetic predisposition and permissive environments to express obesity [adapted from ref. 36].

Arizona have significantly higher levels of both [6]. Obesity is around 10 times more prevalent in Arizona Pima men and 3 times more prevalent in Arizona Pima women than in Mexican Pima; T2DM is four times more prevalent in Arizona Pima than Mexican Pima. Despite considerable genetic similarity, the two Pima groups clearly differ in lifestyle and environment. Whereas the traditional Pima diet was low in saturated fats and high in complex carbohydrates [7]; the typical modern US diet is high in fat and sucrose, which promote weight gain and contributes to glucose intolerance [8]. Similarly the transition away from traditional farming methods has reduced physical activity among Arizona Pima, thus the combination of the US American diet and lower energy expenditure has exacerbated an existing vulnerability to both obesity and T2DM.

The example of the Pima illustrates both the power of genotype to determine phenotype and the extent to which this power is tempered by the prevailing food environment. In contrast to the Arizona Pima population, the average BMI in most Western populations has increased gradually and steadily [3]. In addition, alongside the tendency for BMI to drift upwards, there is an asymmetry at the high end of the distribution curve: BMI >35. Thus, the prevalence of morbid obesity has increased and continues to rise. This suggests considerable individual variation in responsiveness to obesogenic environments.

Individual Differences: Genotype

Given that not all individuals living in so-called 'obesogenic' environments become obese, it is clear that some individuals defend against overweight whilst others demonstrate a propensity to gain weight, perhaps as a function of genotype (fig. 1). Several single gene mutations have been linked to severe obesity [9]. However monogenic disorders represent only a small fraction of population-level obesity. Most obesity is polygenic, involving complex gene-gene and gene-environment interactions. Candidate gene variants predisposing to polygenic obesity generally relate to the regulation of energy intake and expenditure and include the adrenergic receptors, uncoupling proteins, peroxisome proliferator-activated receptor (*PPARG*), proopiomelanocortin (POMC), melanocortin 4 receptor (MC4R) and fat mass and obesity associated (*FTO*) genes [9]. Evidence of gene variants involved in addiction and reward behaviour, such as the neurexin 3 gene (*NRXN3*) predicting greater waist circumference, BMI and obesity [10], is also emerging.

Common gene variants which predict obesity may be detectable through differences in energy expenditure and metabolism, and/or overconsumption. A taxonomy of five potential genotypes has been devised by Bouchard [11], including a thrifty genotype (low metabolic rate and thermogenesis), hyperphagic genotype (poor regulation of appetite, tendency to overeat), a sedens genotype (propensity to be physically inactive), a low lipid oxidation genotype, and an adipogenesis genotype (tendency to store lipids and expand adipocytes). The hyperphagic genotype driving individuals to overconsume in permissive environments with poor control of appetite and satiety is both intuitively appealing and can be subjected to empirical scrutiny. However, characterising this genotype is complex and requires identification of candidate gene variants.

Candidate Gene Variants in Obesity Risk – The Case of *FTO*

Variation in the *FTO* gene has been associated with obesity in genome-wide association studies [12]. It is estimated that adults homozygous for the risk allele (single nucleotide polymorphism (SNP) rs9939609 leading to the AA or AT genotype (at that locus) weigh 3 kg more and are more likely to be obese than those who do not carry the risk allele. The additional weight predicted by this allele is expressed as greater fat mass, thus increasing interest in how this gene may influence food intake and energy expenditure. Frequency of the risk allele is estimated at ~39% in populations of European descent and is strongly associated with BMI-dependent T2DM and obesity in both children and adults [12]. Also in populations of European descent, the combined effects of variants in *FTO* and *MC4R*, described below, are additive. Both predict obesity and T2DM, although the effects of *FTO* may be attenuated by physical activity levels and *MC4R* by sex [13]. Individuals carrying risk alleles of both *FTO* and MC4R weigh almost 4 kg more than those without these risk alleles [14] (fig. 2).

The *FTO* gene is located on chromosome 16 and is expressed mostly in brain, pancreatic islets, adipose tissue and adrenal glands [12]. The expression of *FTO* specifically in hypothalamus, pituitary and adrenal glands, suggests a pivotal role of this gene in the hypothalamic-pituitary-adrenal (HPA) axis and, therefore, potentially the control of eating. Expression of *FTO* is regulated by fasting and re-feeding in the arcuate nucleus of mice [15], supporting its role in energy balance. The SNP rs9939609 resides in the first intron of this gene. Understanding the functional basis of the

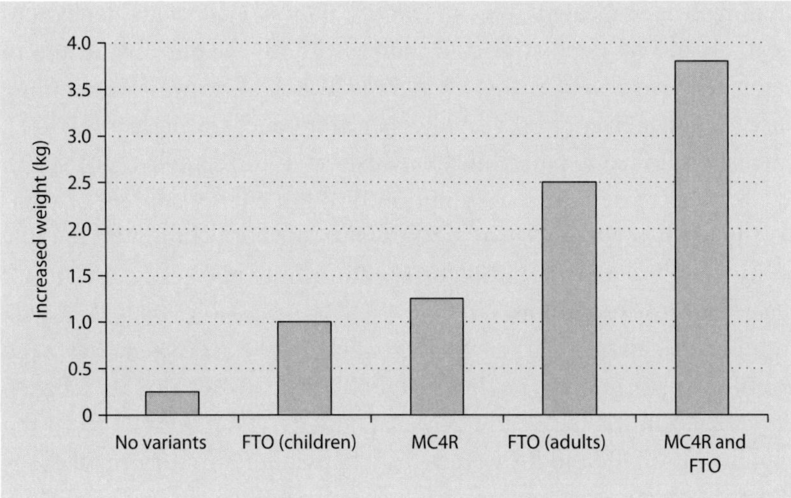

Fig. 2. Additive effect on body weight of carrying risk alleles of either FTO or MC4R [data from refs 12–14, 21].

common variants of the *FTO* gene has involved exploration of the *FTO* genotype in relation to specific phenotypic variation. Compared to wild types, *fto* knockout mice [16] experienced postnatal growth retardation with reduced adipose tissue and lean mass, and expended more energy, but had less spontaneous locomotor activity, and relative hyperphagia. According to this model, *FTO* is functionally involved in energy balance through energy expenditure rather than intake.

In contrast, studies of adults and children carrying the A allele of the *FTO* gene indicate a higher risk for obesity that is conferred by differences in energy intake rather than expenditure. Thus, adults carrying the A allele consumed significantly more energy recorded using 7-day weighed intake than those who do not carry the A allele [17], but did not differ in basal metabolic rate. In children, the A allele was associated with lower satiety responsiveness [18], reported by parents through the Child Eating Behaviour Questionnaire [19]. Using a measure of food intake when children have already been fed (eating in the absence of hunger: EAH), the A allele was associated with higher measured EAH intakes but had no impact on reported activity measures [20].

A direct measure of food intake provides a well-controlled method to examine the eating behaviours of children. Meal intake following water or a low- or a high-energy preload provides indications of satiety responsiveness and of how well children regulate energy intake in the short-term [21]. In response to these preloads, intake in children with the *FTO* risk allele was significantly higher than that of children without, independent of body weight. In the low energy preload condition a difference in meal intake of more than 400 kJ between the genotypic groups was measured [21]. Intake differences of this magnitude repeated over time if not balanced against greater expenditure could lead to greater fat deposition and increased body weight. Inspection of the food type being selected revealed that carriers of the A allele were consuming more energy-dense food, which could reflect a higher preference for these foods [21]. Because energy density is linked to palatability, however, it is unclear whether the risk allele predisposes to an increased overconsumption through energy content or pleasure. In any case, the tendency to eat more (even when sated), and to preferentially select foods high in energy density could plausibly contribute to obesity risk, if not countered by greater energy expenditure.

Energy expenditure in children carrying the A allele of the *FTO* gene has been measured using indirect calorimetry (resting metabolic rate: RMR), doubly labelled water (total energy expenditure: TEE) and physical activity (PA), derived from the difference between the two. TEE and PA were significantly higher in carriers of the A allele but RMR did not differ by genotype when adjusted for body mass. In conclusion, carriers of the risk allele have higher TEE primarily due to greater physical activity levels [21]. Indeed, in a study of more than 8,000 Danish adults [22] homozygotes for the *FTO* risk allele were significantly heavier if they were physically inactive whereas physically active homozygotes had the same BMI as the non-carriers of the *FTO* risk allele. Physical activity serves to attenuate the underlying genetic susceptibility to overweight and obesity conferred by the A allele of *FTO*. Data collected from 2,275 families enrolled in the Avon Longitudinal Study of Parents and Children (ALSPAC) revealed that dietary energy density (DED) derived from 3-day diet records taken at age 10 years and *FTO* contributed independently to greater fat mass by age 13 years [23]. The A allele of *FTO* was associated with more fat mass (0.35 kg for each A allele) and with 0.16 kg more fat mass at age 13 years for each unit of DED (kJ/g). The ALSPAC data has also revealed that children carrying the A allele of the *FTO* gene tend to eat more energy and specifically to consume diets high in fat [24].

Carrying the A allele increases the risk of obesity but this risk can be modified either by physical activity or by consuming a diet low in energy density. The genetic vulnerability to obesity may be expressed in specific eating and activity patterns that are heritable, but only expressed under specific permissive conditions, or acquired through exposure to these conditions.

Individual Differences in Susceptibility to Obesity

Behavioural evidence from experimental studies tends to focus on average energy intake, mean energy expenditure and average scores on measures of appetite regulation. This will mask individual differences in responsiveness to various permissive events or environments. Features of the environment that promote overconsumption are well documented; for example social context, portion size, variety, palatability can all stimulate food intake [25–27]. What is not understood is how individual differences temper the effects of the obesogenic environment in relation to body weight. Indeed, given the obesogenic nature of an environment which encourages overconsumption and discourages physical activity, it is perhaps surprising that only one third of populations in Europe or the US are overweight. Evidently, individual differences in response to these environments serve to promote or inhibit the tendency to gain weight.

Behavioural science offers a variety of methods to examine stable behavioural phenotypes that confer risk, such as food responsiveness using fMRI [28], the reinforcing value of food [29–31], eating in the absence of hunger [32], habitual consumption of high fat diets [33], weak satiety [18, 21], and disinhibition [34]. Food responsiveness could be a key attribute predicting obesity risk. If consumers display an exaggerated response to the presence of food, this places them at considerable risk of weight gain if this tendency is stable in an environment which supports frequent, unnecessary eating. Eating in the absence of hunger captures opportunistic eating without an obvious need to eat and likely reflects the salience of food stimuli in the environment. For others, food is not only salient but has a high reinforcement value, such that consumers are prepared to work hard to obtain a food reward above other types of reward [31]. This latter form of food responsiveness is higher in children who are overweight [29, 31]. Also the relative reinforcing value of food gauged by questionnaire predicts weight gain after 1 year in 7- to 10-year-old children [30]. Thus, the tendency to over-respond to the presence of food cues as characterised by Lowe et al. [35] in the 'power of food' scale or to work to obtain food [31] captures a stable, disposition and one which appears to predict weight gain [30].

Taken together, sophisticated methods to genotype individuals along with advances in behavioural phenotyping characterise specific gene-environment interactions that predict obesity. There are few reliable options available to change permissive environments and fewer still to alter the human genome, however, at the interface between inheritance and environment lies behaviour, with the potential, at least, for change.

References

1 Lobstein T, Baur L, Uauy R, IASO International Obesity TaskForce: Obesity in children and young people: a crisis in public health. Obesity Rev 2004; 5(suppl 1):4–104.

2 World Health Organisation (WHO): Diet, the food supply and obesity in the Pacific. 2003 Report by the WHO Western Pacific Regional Office.

3 World Health Organisation (WHO): Diet, nutrition and the prevention of chronic diseases. Report of a joint WHO/FAO expert consultation. Geneva, 2003, WHO Techl Rep Ser 916.

4 Houghton P: People of the Great Ocean. Cambridge, Cambridge University Press, 1996.

5 Serjeantson SW, Ryan DP, Thompson AR: The colonization of the Pacific: the story according to human leukocyte antigens. Am J Hum Genet 1982;34:904–918.

6 Schulz LO, Bennett PH, Ravussin E, Kidd JR, Kidd KK, Esparza J, Valencia ME: Effects of traditional and western environments on prevalence of type 2 diabetes in Pima Indians in Mexico and the US. Diabetes Care 2006;29:1866–1871.

7 Pratley RE: Gene-environment interactions in the pathogenesis of type 2 diabetes mellitus: lessons learned from the Pima Indians. Proc Nutr Soc 1998; 57:175–81.
8 Swinburn BA, Boyce VL, Bergman RN, Howard BV, Bogardus C: Deterioration in carbohydrate metabolism and lipoprotein changes induced by modern, high fat diet in Pima Indians and Caucasians. J Clin Endocrinol Metab, 1991;73:156–165.
9 Farooqi S, O'Rahilly S: Genetics of obesity in humans. Endocr Rev 2006;27:710–718.
10 Heard-Costa NL, Zillikens MC, Monda KL, et al: NRXN3 is a novel locus for waist circumference: a genome-wide association study from the CHARGE Consortium. PLoS Genet 2009;5:e1000539. Epub 2009 Jun 26.
11 Bouchard C: The biological predisposition to obesity: beyond the thrifty genotype scenario. IJO 2007;31:1337–1339.
12 Frayling TM, Timpson NJ, Weedon MN, et al: A common variant in the FTO gene is associated with body mass index and predisposes to childhood and adult obesity. Science 2007;316:889–894.
13 Cauchi S, Stutzmann F, Cavalcanti-Proença C, et al: Combined effects of MC4R and FTO common genetic variants on obesity in European general populations Mol Med 2009;87:537–546
14 Loos RJ, Lindgren CM, Li S, et al: Common variants near MC4R are associated with fat mass, weight and risk of obesity. Nature Genetics 2008, 40:768–775.
15 Fredriksson R, Hägglund M, Olszewski PK, Stephansson O, Jacobsson JA, Olszewska AM, Levine AS, Lindblom J, Schiöth HB: The obesity gene, FTO, is of ancient origin, up-regulated during food deprivation and expressed in neurons of feeding-related nuclei of the brain. Endocrinology 2008;149:2062–2071
16 Fischer J, Koch L, Emmerling C, Vierkotten J, Peters T, Brüning JC, Rüther U: Inactivation of the Fto gene protects from obesity. Nature 2009;458:894–898.
17 Speakman J, Jackson DM, Johnstone AM: Polymorphisms of the FTO gene are associated with variation in energy intake, but not energy expenditure. Obesity 2008;16:1961–1965.
18 Wardle J, Carnell S, Haworth CM, Farooqi IS, O'Rahilly S, Plomin R: Obesity associated genetic variation in FTO is associated with diminished satiety J Clin Endocrinol Metab 2008;93:3640–3643.
19 Wardle J, Guthrie CA, Sanderson S, Rapoport, L: Development of the children's eating behaviour questionnaire. J Child Psychol Psychiatry 2001;42: 963–970.
20 Wardle J, Llewellyn C, Sanderson S, Plomin R. The FTO gene and measured food intake in children. Int J Obes 2009;33:42–45.
21 Cecil JE, Tavendale R, Watt P, Hetherington MM, Palmer CN: An obesity-associated FTO gene variant and increased energy intake in children. N Engl J Med 2008;359:2558–2566.
22 Andreasen CH, Stender-Petersen KL, Mogensen MS, et al: Low physical activity accentuates the effect of the FTO rs9939609 polymorphism on body fat accumulation. Diabetes 2008;57:264–268.
23 Johnson L, van Jaarsveld CH, Emmett PM, Rogers IS, Ness AR, Hattersley AT, Timpson NJ, Smith GD, Jebb SA: Dietary energy density affects fat mass in early adolescence and is not modified by FTO variants. PLoS One 2009;4:e4594.
24 Timpson NJ, Emmett PM, Frayling TM, Rogers I, Hattersley AT, McCarthy MI, Davey Smith G: The fat mass- and obesity-associated locus and dietary intake in children Am J Clin Nutr 2008;88:971–978.
25 Hetherington MM: Cues to overeat: psychological factors influencing over-consumption. Proc Nutr Soc 2007;66:113–123.
26 Rolls BJ, Roe LS, Meengs JS: The effect of large portion sizes on energy intake is sustained for 11 days. Obesity 2007;15:1535–1543.
27 Yeomans MR: Palatability and the micro-structure of feeding in humans: the appetizer effect. Appetite 1996;27:119–133.
28 Stice E, Spoor S, Bohon C, Small DM: Relation between obesity and blunted striatal response to food is moderated by TaqIA A1 allele. Science 2008; 322:449–452.
29 Epstein LH, Robinson JL, Temple JL, Roemmich JN, Marusewski AL, Nadbrzuch RL: Variety influences habituation of motivated behavior for food and energy intake in children. Am J Clin Nutr 2009;89: 746–754.
30 Hill C, Saxton J, Webber L, Blundell J, Wardle J: The relative reinforcing value of food predicts weight gain in a longitudinal study of 7–10-y-old children. Am J Clin Nutr 2009;90:1–6.
31 Temple JL, Legierski CM, Giacomelli AM, Salvy SJ, Epstein LH: Overweight children find food more reinforcing and consume more energy than do nonoverweight children. Am J Clin Nutr, 2008;87: 1121–1127.
32 Fisher JO, Birch LL: Eating in the absence of hunger and overweight in girls from 5 to 7 y of age. Am J Clin Nutr 2002;76:226–231.

33 Blundell JE, Cooling J: High-fat and low-fat (behavioural) phenotypes: biology or environment? Proc Nutr Soc 1999;58:773–777.
34 Bryant EJ, King NA, Blundell JE: Disinhibition: its effects on appetite and weight regulation. Obes Rev 2008;9:409–419.
35 Lowe MR, Butryn ML, Didie ER, Annunziato RA, Thomas JG, Crerand CE, Ochner CN, Coletta MC, Bellace D, Wallaert M, Halford J: The Power of Food Scale: a new measure of the psychological influence of the food environment. Appetite 2009;53:114–118.
36 Loos RJ, Bouchard C: Obesity – is it a genetic disorder? J Internal Medicine 2003;254:401–425.

Marion M. Hetherington
Institute of Psychological Sciences, University of Leeds
LS2 9JT, Leeds (UK)
Tel. +44 0 113 3438472, Fax +44 0 113 3435749, E-Mail M.Hetherington@Leeds.ac.uk

Author Index

Subject Index